Energy Co[nsumption in Bu]ildings

A Guide to Part L of the
Building Regulations

Energy Conservation in Buildings

A Guide to Part L of the Building Regulations

J.R. Waters
BSc, MPhil, PhD, MCIBSE, CEng

Blackwell
Publishing

© 2003 J.R. Waters
Blackwell Publishing Ltd
Editorial Offices:
9600 Garsington Road, Oxford OX4 2DQ, UK
 Tel: +44 (0)1865 776868
108 Cowley Road, Oxford OX4 1JF, UK
 Tel: +44 (0)1865 791100
Blackwell Publishing USA, 350 Main Street, Malden,
MA 02148-5018, USA
 Tel: +1 781 388 8250
Iowa State Press, a Blackwell Publishing Company,
2121 State Avenue, Ames, Iowa 50014-8300, USA
 Tel: +1 515 292 0140
Blackwell Munksgaard, 1 Rosenørns Allé, P.O. Box
227, DK-1502 Copenhagen V, Denmark
 Tel: +45 77 33 33 33
Blackwell Publishing Asia Pty Ltd, 550 Swanston
Street, Carlton South, Victoria 3053, Australia
 Tel: +61 (0)3 9347 0300
Blackwell Verlag, Kurfürstendamm 57, 10707 Berlin,
Germany
 Tel: +49 (0)30 32 79 060
Blackwell Publishing, 10 rue Casimir Delavigne, 75006
Paris, France
 Tel: +33 1 53 10 33 10

First published 2003

A catalogue record for this title is available from the
British Library

ISBN 1-4051-1253-0

Library of Congress
Cataloging-in-Publication Data
is available

Set in 10/12.5pt Times
by DP Photosetting, Aylesbury, Bucks
Printed and bound in Great Britain by
TJ International, Padstow, Cornwall

For further information on
Blackwell Publishing, visit our website:
www.blackwellpublishing.com

Contents

Preface

Pollution of the atmosphere by carbon dioxide emissions and the resulting rise in global temperature is posing a threat to the global environment to the extent that it is now a matter of international concern. It is recognised by governments as well as scientific authorities that action to counter this threat is both necessary and urgent. Although there are many sources of carbon emissions into the atmosphere, buildings and their services systems are among the most significant, and all governments have a duty to take action. Consequently, the conservation of fuel and power in buildings is an essential part of the UK government's strategy to reduce national energy consumption, and hence to reduce carbon emissions into the atmosphere.

The revision to Part L of the Building Regulations which came into force on 1 April 2002 provides the legislative mechanism putting this strategy into effect. The designers, constructors and operators of buildings are required not only to conserve energy but also to be aware of the implications of their actions for carbon emissions. Indeed, many of the criteria in the new legislation are expressed directly in terms of the amount of carbon released by buildings into the atmosphere.

There are several important features of the new Part L. Unlike previous versions, it covers almost all buildings and almost all potential causes of building energy consumption, and it includes many new features, of which air tightness standards are a particular example.

Furthermore, the new Part L places an obligation on everyone, including architects and designers, contractors and sub-contractors, and building operators, to contribute to the control and reduction of energy consumption. Consequently the guidance provided in the Approved Documents L1 and L2 is detailed and extensive. It is therefore vital that all concerned should understand both the philosophy of the regulations and be fully conversant with the detailed mechanisms of its application. This book has been written as an essential companion to both of the Approved Documents, L1 and L2, describing, explaining and expanding on the information and guidance they contain.

An important aspect of Part L is the understanding of its application to particular cases, and so this book contains numerous worked examples that reinforce, and in some cases go beyond, those that appear in the Approved Documents themselves.

The airtightness of buildings is a new feature in the regulatory framework, and is a topic with which many building professionals are unfamiliar. In recognition of this, a chapter is devoted to explaining its background and principles, and the procedures required for air leakage testing. Similarly, there is now a strong requirement to avoid excessive thermal bridging, and a chapter is devoted to this.

It is hoped that by assisting all those professionals who are involved in the design, construction, operation and regulation of the building process, including architects, builders, building services engineers, building control officers, building operators and facility managers, this book will make a contribution to the larger objective of controlling carbon emissions and global warming.

Acknowledgements

I am grateful to the following for permission to reproduce copyright material:

BRE/CRC for permission to use table 3 from IP17/01 and data from tables 2, 3 and 4 from BRE Digest 457; BRECSU for permission to use data from tables 3.1, 3.2, 3.3 and 3.4 from SAP 2001; the British Standards Institution for permission to reproduce tables from BS EN ISO 6946 under licence number 2002SK/0365.

Information from the tables in the Approved Documents L1 and L2 is crown copyright and reproduced with the permission of the Controller of HMSO and the Queen's Printer for Scotland.

BRE/CRC and BRECSU publications, plus most others, may be obtained from: BRE Bookshop, 151 Rosebery Avenue, London EC1R 4GB. Tel: 020 7505 6622. website: www.brebookshop.com email: brebookshop@emap.com

BSI publications can be obtained from BSI Customer Services, 389 Chiswick High Road, London W4 4AL. Tel: 020 8996 9001. email: cservices@bsi-global.com

I would also like to thank my colleague Dr Martin Simons for kindly reading and commenting on the draft.

1 Introduction

National Building Regulation Standards for the insulation of buildings were first introduced in 1965, and at that time were applicable only to dwellings. Their objective was part of an overall aim to maintain minimum standards of health and safety within buildings. Since then, the insulation standards have been progressively improved and their scope extended to all buildings. Simultaneously, the objective of this part of the Building Regulations has expanded from a concern for health and safety within individual buildings to include the conservation of fuel and power at a national level. Originally, the trend towards conservation of fuel and power was a defensive reaction to instability in the price and availability of oil on the world market, as well as being a contribution to environmental welfare. More recently, steadily increasing pollution of the atmosphere by various gases has become a major international concern. In this context, the most relevant gas is carbon dioxide, and for several reasons:

- Increases in carbon dioxide levels in the atmosphere lead to global warming and climate change
- Carbon dioxide levels in the atmosphere are on a rising trend
- Carbon dioxide is the main end product of the consumption of fuel for energy
- As buildings are, in total, among the largest consumers of fuel and energy, they are also one of the largest contributors to the increase in atmospheric carbon dioxide and hence to global warming and climate change.

The average global temperature is the result of a balance between the short-wave solar radiation penetrating the earth's atmosphere and the outgoing thermal long-wave radiation. Several gases in the earth's atmosphere, most notably carbon dioxide, absorb the outgoing radiation more efficiently than the incoming radiation, hence maintaining the average temperature at or about its customary level. The connection between atmospheric carbon dioxide and global temperature can be seen in Fig. 1.1, which is a simplified graph of the changes which are believed to have occurred naturally over time. This shows global temperature closely following carbon dioxide levels in a repeating cyclic fluctuation. If this fluctuation were to continue, carbon dioxide and temperature levels would, at the present time, be expected to be near peak values and be beginning to fall. However, both are continuing to rise to levels that are much higher than ever before. Although previous peaks have been around 280 ppm

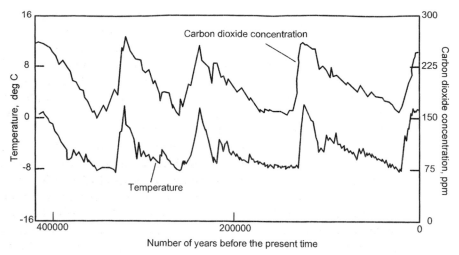

Fig. 1.1   Carbon dioxide concentration and global temperature.

(parts per million), by the year 2000 the global CO_2 concentration was on a sharply rising trend, and had reached 360 ppm. In the wake of this rise, global temperature has been found to be rising at a rate, which is expected to continue, of about 1° Celsius every 40 years. The consequences of allowing this rise to continue are thought to be unacceptable, and so action to reduce or even prevent it is considered necessary.

Clearly, controlling carbon dioxide levels and global warming is an international problem, which requires action on several fronts. As one part of its contribution to that process, the UK government has made the relevant part of the Building Regulations more stringent, and has emphasised that the objective of conserving fuel and power is to reduce carbon emissions into the atmosphere. This has been done by using carbon emissions as one of the principal criteria by which the performance of a building is judged. This in turn has had a substantial impact on the structure and detail of the Regulations. For carbon emission targets to be met, all aspects of a building's design and operation that might affect those emissions must be considered and are therefore included in the regulatory framework. An alternative method of rating the energy performance of a building is to convert its on-site consumption to an equivalent amount of primary energy. The primary energy figure is a measure of the total amount by which a natural resource has been depleted for every unit of useable energy delivered to an end user. The primary energy, therefore, takes account of all the extra energy used in producing the fuel and delivering it to the building in a form that is ready for consumption. Some of the literature relevant to Part L of the Building Regulations refers to primary energy rather than carbon emissions. In practice, the two measures are performing the same function, and the carbon emission factors given in Table 3.5 for various fuels are similar, apart from a conversion factor, to primary energy ratios.

As far as the Building Regulations are concerned, requirements for the conservation of fuel and power are dealt with by Part L of Schedule 1, and there are two Approved Documents providing practical guidance for meeting the requirements. The first, Approved Document L1 (AD L1), is applicable to dwellings, and the second, Approved Document L2 (AD L2), is applicable to all buildings other than dwellings. The two documents are separate publications and are substantially different. However, there are some common elements between them and these areas of commonality are listed in Table 1.1. All other material differs between the two documents, even though there is some similarity in the wording in places. It is therefore important to maintain the distinction between the regulations and requirements of L1 and L2 when applying them to a specific building project. Consequently, in this book parts L1 and L2 are treated separately. However, to avoid undue repetition, the common material in 'Use of Guidance' and in 'Introduction to the Provisions' (L1 paragraphs 0.5 to 0.16; L2 paragraphs 0.9 to 0.20) is considered first, then L1 and L2 are described separately in Chapters 2 and 3. The common material in Appendices A, B, C and D is presented in Chapters 4, 5, 6 and 7, followed by the remaining Appendices of L1 and L2 in Chapters 10, 11, 12 and 13. The requirement to achieve an acceptable standard of airtightness for the external

Table 1.1 Material common to L1 and L2.

Section	Paragraph in L1	Paragraph in L2	Comment
Use of Guidance			Identical except for 'Mixed Use Development' in L2
Technical risk	0.5	0.9	
Thermal conductivity and transmittance	0.6 to 0.9	0.10 to 0.13	
U-value reference tables	0.10	0.14	
Calculation of U-values	0.11 to 0.13	0.15 to 0.17	
Roof window	0.14	0.18	
Basis for calculating areas	0.15	0.19	
Air permeability	0.16	0.20	
Limiting thermal bridging at junctions and around openings	1.30 to 1.32	1.90 to 1.11	Identical except for some additional references in L2
Appendix A: Tables of U-values	The whole Appendix	The whole Appendix	
Appendix B: Calculating U-values	The whole Appendix	The whole Appendix	
Appendix C: U-values of ground floors	The whole Appendix	The whole Appendix	
Appendix D: Determining U-values for glazing	The whole Appendix	The whole Appendix	The principle is the same but the examples differ

fabric of a building, and the possible need to test for air leakage, is a new development and is discussed in Chapter 14.

1.1   Use of guidance

Although this section is common to both Approved Documents, there are two points of note. The first point arises from the fact that the thermal performance of a building is the result of a complex interaction between many different systems (building fabric, services systems, fuel supply, availability of ambient energy, solar gains, building usage, etc.). It is therefore possible that novel design solutions can be found which cannot easily be assessed for compliance by the guidance and methods contained in the Approved Documents, but which nevertheless can be shown to satisfy the requirements for the conservation of fuel and power by some other means. In any case, it is not legally mandatory to use the Approved Documents, provided that it can be demonstrated, to the satisfaction of a Building Control Body, that the legal requirement of Part L of Schedule 1 to conserve fuel and power has been met.

The second point of note concerns the statement that Building Regulations do *not* require anything to be done *except* for the purpose of securing reasonable standards of health and safety. Parts L1 and L2 (along with Part M and paragraphs H2 and J6) are excluded from this statement because the requirement to conserve fuel and power is *in addition* to health and safety, and not a substitute for it.

1.2   General definitions applicable to L1 and L2

Both L1 and L2 begin with a reminder that designing to minimise energy consumption may carry the risk of technical problems in other areas. High levels of thermal insulation, and careful attention to draught proofing can cause problems due to:

- Interstitial condensation
- Surface condensation in roof spaces
- Inadequate ventilation for occupants
- Inadequate ventilation and air supply to combustion systems and flues.

Other potential problems are:

- Rain penetration causing, among other things, damage to thermal insulation. This is often associated with failure of flat roof coverings due to the high levels of thermal stress induced by high levels of thermal insulation within the roof construction.
- Sound transmission, due to the fact that materials which provide good

thermal insulation are normally very poor at providing insulation against the transmission of sound.

Guidance on the avoidance of these related technical risks must be found elsewhere, and a number of sources of information are suggested:

- BRE Report No. 262: *Thermal Insulation: avoiding risks*, 2002 edition [1]
- Approved Document F, *Ventilation*
- Approved Document J, *Combustion appliances and fuel storage systems*
- Approved Document E, *Resistance to the passage of sound*
- *Limiting thermal bridging and air leakage: Robust construction details for dwellings and similar buildings* [2].

Both L1 and L2 also contain a series of definitions, as follows, that apply throughout the documents.

Thermal conductivity (the λ-value)
This is the rate at which heat will pass through unit area of a material when there is unit temperature gradient across the material. It is usually expressed in watts per square metre for a temperature gradient of one degree Kelvin per metre, and is given the symbol λ (Greek lambda). The units of λ are thus $\frac{W}{m^2}/\frac{K}{m}$, which simplifies to $\frac{W}{m.K}$.

Thermal transmittance (the U-value)
This is the rate at which heat will pass through unit area of a material (or a construction made up of several materials) when there is unit temperature difference between the environments on the opposite sides of the material. It is usually expressed in watts per square metre for a temperature difference of one degree Kelvin, and is given the symbol U. The units of U are thus $\frac{W}{m^2.K}$.

If measured test results for λ and/or U-values are available, these should be used. Manufacturers are often able to provide such information for the λ-values of their materials, but where these are not available, values may be obtained from data published in Appendix A of the Approved Documents or in any other authoritative publication (e.g. BS EN 12524 [3] or CIBSE Guide, section A3 [4]). The measurement standards that should be followed are BS EN 12664 [5], BS EN 12667 [6] and BS EN 12939 [7]. Manufacturers sometimes also provide test results for the U-values of various construction elements that incorporate their materials, the relevant measurement standards being BS EN ISO 8990 [8] or, for windows and doors, BS EN ISO 12567-1 [9]. More usually, the U-values supplied by manufacturers are obtained by calculation from λ-values. Where calculated values are used, care must be taken to ensure that proper allowance has been made for thermal bridging effects. These are most likely to arise when the construction includes joists, structural or other types of framing, or any material or component that breaks through an insulation layer. The bridging effect can only be ignored when the bridged material and the bridging material have sufficiently similar thermal properties (i.e. if the difference in their thermal resistances is less than 0.1 $\frac{m^2 K}{W}$). This will normally apply to mortar joints in brickwork, but not necessarily to mortar joints in lightweight blockwork. Calculation procedures for U-values are specified in:

- BE EN ISO 13789 [10] and BRE/CRC [11] for calculation methods and conventions
- BS EN ISO 6946 [12] for walls and roofs
- BS EN ISO 13370 [13] for ground floors
- BS EN ISO 10077-1 [14] or prEN ISO 10077-2 [15] for windows and doors
- BS EN ISO 13370 [13] or BCA/NHBC Approved Document [16] for basements
- BS EN ISO 10211-1 [17] and BS EN ISO 10211-2 [18] for thermal bridges.

Appendix B of AD L1 and L2 provides a simplified method, based on BS EN ISO 6946 [12], which is suitable for the calculation of the U-values of most wall and roof constructions. Chapter 5 gives example calculations.

Exposed element
This means an element exposed to the outside air, and includes:

- A suspended floor over a ventilated or unventilated void
- An element exposed to the outside air indirectly via an unheated space
- An element in a floor or basement that is in contact with the ground.

It should be noted that an element exposed to the outside air indirectly via an unheated space was previously known as a 'semi-exposed' element. The calculation of the U-value of such an element must now use the method given in SAP 2001 [19]. It should also be noted that a wall separating a dwelling from any other premises that are heated to the same temperature does not require thermal insulation.

Roof window
This is defined as a window in the plane of a pitched roof. Within AD L1 and L2 such a window may be considered as a rooflight.

Basis for calculating areas
When evaluating areas, measurements should be taken on the internal faces of external elements. The areas of projecting bays must be included. Roof areas should be measured in the same plane as the roof insulation, and floor areas should include non-useable space (e.g. stairwells and builders' ducts).

Air permeability
This is a measure of the leakiness of the building fabric to unwanted internal to external air exchange. It is defined as the average volume of air, per unit area,

which passes through the building envelope, when there is an internal to external pressure difference. It is normally expressed in cubic metres per hour, per square metre of building envelope area, at a pressure difference of 50 Pa. Within AD L1 and L2, the envelope area is taken to be the total area of walls, floor and roof separating the interior volume from the external environment. Air permeability does not include deliberate leakage paths between the inside and the outside, such as flues, ventilation ducts, air bricks, etc. These are sealed up during any measurement procedure, so that only cracks, gaps at joints and similar leakage paths are included. The lower the air permeability the better, as this shows that the building fabric is more airtight.

Standard assessment procedure (SAP)

This is the UK government's procedure for rating the energy cost performance of dwellings [19], and is defined in Chapter 9. Separately and independently of Part L1, Building Regulation 16 requires that whenever a new dwelling is created (either by building work or by a material change of use), the energy rating of that dwelling must be calculated by means of the standard assessment procedure. The result of the SAP calculation cannot be used by itself to demonstrate compliance with Part L1. However, the SAP calculation can be continued to find the carbon index of the dwelling. If the carbon index is above a specified minimum, compliance has been demonstrated (see section 2.3.4).

1.3 Testing

Building Regulation 18 has been extended. Previously it gave local authorities the power to test drains and private sewers for compliance with Part H. Now, Regulation 18 allows a local authority to make such tests as may be necessary:

- For compliance with Building Regulation 7, which specifies that all building work should be carried out with proper materials and in a workmanlike manner, *and also*
- For compliance with any of the applicable requirements of Schedule 1.

The inclusion of Schedule 1 within Regulation 7 confers very wide powers on a local authority to require testing. Much of that testing could be done beforehand on components, using approved test procedures, and the results provided in the form of appropriate certificates. However, there are several areas where testing can only be carried out and be effective when the building is complete, and two of these are of particular significance to Part L:

- Airtightness testing of buildings to test for compliance with the air permeability criterion
- Testing for continuity of insulation and the avoidance of thermal bridging.

In the case of airtightness testing, the Approved Documents point to only one type of test, the fan pressurisation test, and the criterion for compliance is written specifically in terms of the result which this test provides. Indeed, for large buildings above 1000 m² floor area, Approved Document L2 appears to offer no alternative to the test as a means of demonstrating compliance.

For the testing of the continuity of insulation, there is a similar lack of flexibility. Unless an authoritative certificate can be provided stating that the design details and building techniques are appropriate, the Approved Documents require satisfactory results from a test of the whole of the visible external envelope using infrared thermography.

2 The Conservation of Fuel and Power in Dwellings

Part L1 of Schedule 1 to the 2001 Regulations is concerned with the conservation of fuel and power in dwellings. This is supported by Approved Document L1 which covers the following topics:

- Compliance by the elemental method
- Compliance by the target U-value method
- Compliance by the carbon index method
- Limiting thermal bridging and air leakage
- Space heating controls and hot water systems
- Commissioning, operating and maintenance of heating and hot water systems
- Insulation of pipes and ducts
- External lighting
- Conservatories.

The purpose of Part L1 is to minimise the environmental impact of dwellings with respect to the depletion of fuel reserves and the degradation of the atmosphere by carbon dioxide emissions.

2.1 The legal requirement for the conservation of fuel and power in dwellings

The legal regulation requires that a dwelling should be so designed and constructed as to make reasonable provision for the conservation of fuel and power, and that this shall be achieved by attention to all of the following four points:

- Limiting heat loss through the fabric of the building, from hot water pipes, from air ducts used for space heating and from hot water vessels
- Ensuring that the space heating and hot water systems which are provided are energy efficient
- Ensuring that internal and external lighting systems are designed to use energy efficiently
- Ensuring that, by the provision of sufficient information, building occupiers are able to operate and maintain the heating and hot water services efficiently, so as to use no more energy than is reasonable in the circumstances.

9

In addition, under Regulation 16, all new dwellings (whether created by new construction or by change of use) are required to be given a SAP Energy Rating. This must be supplied to the local authority or approved inspector.

2.2   AD L1 – Section 0 General Guidance

Approved Document L1 gives general and specific guidance on the energy efficient measures which will satisfy the requirements. These include not only means to limit heat loss from the building and its heating and hot water services, but also means to limit the energy consumed by the heating, hot water and lighting equipment.

Heat losses through walls, roofs, floors, windows and doors, etc. must be kept within acceptable limits by means of suitable levels of thermal insulation. Also, where it is possible and appropriate, any beneficial gains from solar heat and more efficient heating systems should be taken into account.

Heat may also be lost by unplanned air leakage through the fabric of a building, particularly around doors, windows, and junctions between elements or components. This requires attention not only to draught stripping of doors and windows, but also consideration of potential air leakage paths through all other elements of the building.

Heat losses from hot water pipes, hot air ducts, hot water vessels, and all their connections must also be controlled by the application of sufficient insulation. If, however, a particular component contributes to the space heating in an efficient manner, the insulation may be omitted.

The energy efficiency of the systems that provide space heating and hot water has two aspects. Firstly, the heat generating device (i.e. the boiler) must extract heat from the fuel in an efficient manner. This means that it must have a sufficiently high combustion efficiency. Secondly, the heat must be delivered and distributed efficiently. This means that the equipment must have controls for space and water temperatures and for the timing of its operation. Furthermore, the systems must have been appropriately commissioned and be capable of being operated in an efficient manner by the user.

The requirements for lighting systems apply both to internal lighting and to external lighting fixed to the building. All lighting systems should, where appropriate, be fitted with energy efficient lamps. In addition, external lighting systems should have a control system, either fully automatic or combined manual/automatic, which acts in such a way as to conserve energy.

The energy efficiency of heating and hot water systems can be affected by the manner in which they are operated. Incorrect usage may result in poor energy performance from an otherwise energy efficient system. Consequently, there is a requirement to provide easily understandable information that will enable the occupier to operate and maintain these systems so that they do not waste energy. This information should also include the results of performance tests on the systems.

2.3   AD L1 – Section 1 Design and Construction

2.3.1   Specific guidance

For specific guidance, Approved Document L1 offers three methods for demonstrating that reasonable provision has been made for limiting heat loss through the building fabric.

The elemental method has the advantages that it involves a minimum of calculation effort, and is appropriate for alterations and extensions as well as for new construction. However, it allows less flexibility in the design of the dwelling than other methods and can only be used with certain heating systems:

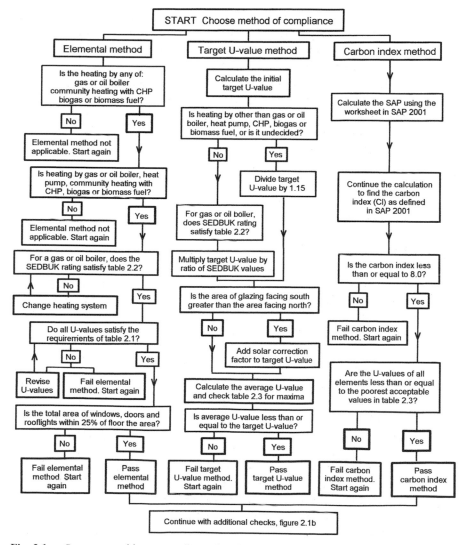

Fig. 2.1a   Summary guide to compliance for dwellings.

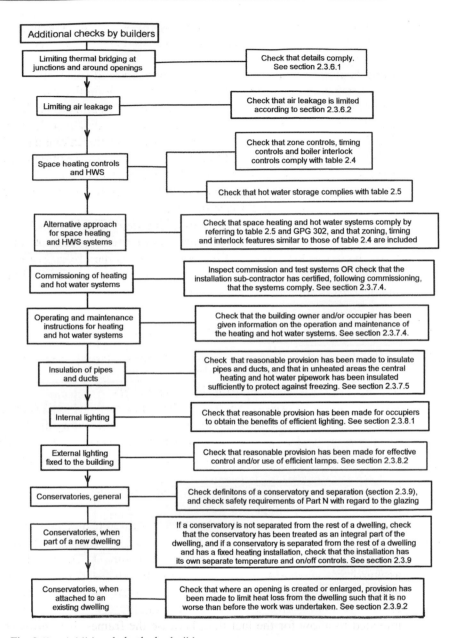

Fig. 2.1b   Additional checks by builders.

- Systems based on an efficient gas or oil boiler
- Community heating with CHP
- Systems based on biogas or biomass fuel.

The elemental method *cannot* be used for dwellings using any other heating system (such as direct electric heating).

The target U-value method can only be used for complete dwellings. However, it can be used with any heating system. Its advantage is that it allows adjustment of the areas of windows, doors and rooflights by taking into account insulation levels, heating system efficiency and the possibility of solar gain.

The carbon index method can also be used for complete dwellings only and with any heating system. It is intended to allow substantial flexibility in design, and is likely to be the most suitable method if the design includes unconventional or novel features. In practice it is based on the calculation of the SAP rating, which is then converted to the carbon index. The requirement is that the carbon index for a dwelling (or for each dwelling in a block of flats or converted building) should be not less than 8.0. Values below 8.0 fail the requirement whereas values of 8.0 and above satisfy it.

The Approved Document provides a 'Summary Guide', which is a comprehensive check-list to help the designer choose the most suitable of the above three methods. This check-list is shown in flow chart form in Fig. 2.1a. The summary guide also indicates the additional checks which should be made by builders; these are shown in Fig. 2.1b.

2.3.2   The elemental method for dwellings

In the elemental method, compliance is demonstrated by specifying:

- Maximum U-values for walls, floors and roofs
- Maximum area-weighted U-values for windows, doors and rooflights
- Maximum combined area of windows, doors and rooflights
- Minimum boiler efficiencies for the heating system boiler (the SEDBUK rating, or Seasonal Efficiency of a Domestic Boiler in the UK).

The maximum U-values allowed by the elemental method are shown in Table 2.1, and are illustrated in Fig. 2.2.

If an element is exposed to the outside through an unheated space (e.g. garage, atrium, etc.) the unheated space can be ignored, provided it is assumed that the element itself is exposed to outside air, i.e. the element is assigned its normal U-value. Alternatively, the unheated space may be included in the U-value calculation for the element, as demonstrated in Chapter 5. The slightly higher permitted U-value for openings in metal frames is intended to allow for the fact that, because the frames themselves are usually more slender than wood or UPVC and therefore have a higher proportion of glazing, there is on average additional solar gain to offset the heat load of the dwelling.

The area weighted U-value of the windows, doors and rooflights, U_{WDR}
The maximum values in Table 2.1 refer to the area weighted average of all the windows, doors and rooflights in the dwelling; this can be calculated from the formula at the top of page 15:

Table 2.1 Maximum U-values for the elemental method.

Exposed element	Maximum U-value, W/m^2K	Comment
Pitched roof, insulation between rafters	0.20	Any part of a roof having a pitch of 70° or more can be considered as a wall.
Pitched roof with integral insulation	0.25	
Pitched roof, insulation between joists	0.16	
Flat roof	0.25	Roof of pitch 10° or less
Walls, including basement walls	0.35	
Floors, including ground floors and basement floors	0.25	
Windows, doors and rooflights, glazing in metal frames, area-weighted average	2.20	Rooflights include roof windows. The higher U-value for metal frames allows for extra solar gain due to greater glazed proportion
Windows, doors and rooflights, glazing in wood or PVC frames, area-weighted average	2.00	

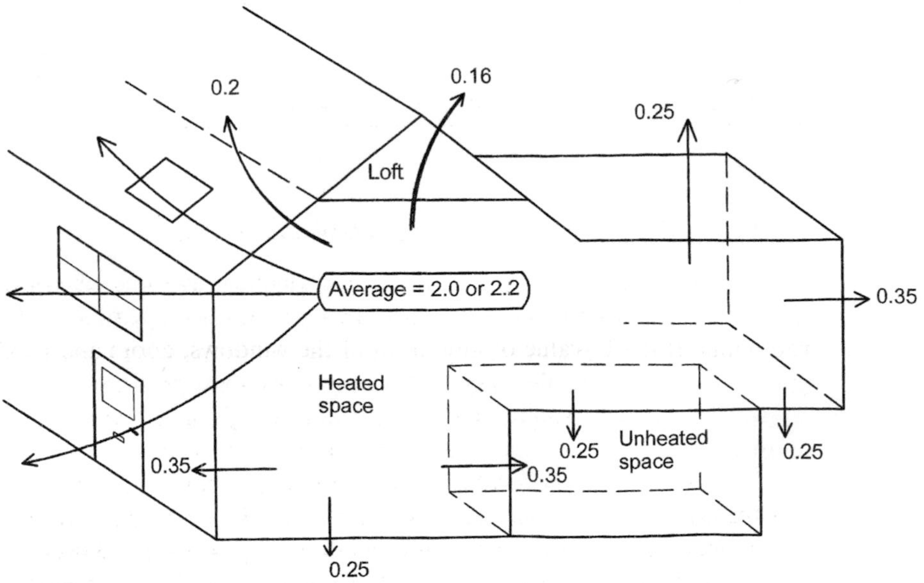

Fig. 2.2 Maximum U-values allowed by the elemental method for dwellings.

$$\text{Average U} - \text{value, } U_{WDR} = \frac{A_W U_W + A_D U_D + A_R U_R}{A_W + A_D + A_R}$$

where A_W, A_D and A_R are, respectively, the areas of windows, doors and rooflights

U_W, U_D and U_R are their U-values.

Maximum areas of windows, doors and rooflights

The elemental method allows a maximum combined area, called the *standard area provision*, which is equal to 25% of the total floor area. If the average U-value, U_{WDR} of the windows, doors and rooflights, does not exceed the value given in Table 2.1, *and also* their combined total area does not exceed 25% of the total floor area, then the requirement is met.

Minimum boiler efficiencies

The minimum boiler efficiencies allowed by the elemental method are shown in Table 2.2. There is no provision to use lower values. The SEDBUK efficiency (also called the SAP seasonal efficiency) of a large number of boilers can be found on the internet (www.SEDBUK.com). If the SEDBUK of the proposed boiler is not available, it is permissible to use the appropriate seasonal efficiency value as given in SAP 2001.

Table 2.2 Minimum boiler seasonal efficiencies (SEDBUK) for the elemental method.

Central heating system fuel	SEDBUK efficiency, %
Mains natural gas	78
LPG	80
Oil	85
Oil-fired combination (as calculated using the SAP-98 method)	82

2.3.2.1 Adjustments allowed by the elemental method

When applying the elemental method to a complete dwelling, the only adjustment available concerns the standard area provision for windows, doors and rooflights. If the U-value of any or all of the windows, doors and rooflights is greater than normal, due say to lower performance glazing, then the area-weighted average U-value may exceed the value given in Table 2.1, and may therefore lead to a failure to meet the requirement. In such a case, the total area may be reduced to compensate. The over-riding principle is that the actual heat loss through the windows, doors and rooflights should be no more than if the maximum average U-value and the standard area provision had been met. If A_F is the total floor area, the maximum allowable heat loss through windows, doors and rooflights may be found from:

- For metal frames: maximum heat loss $= 2.2 \times 0.25A_F = 0.55A_F$
- For all other frames: maximum heat loss $= 2.0 \times 0.25A_F = 0.50A_F$

The actual heat loss is given by:

$$\text{Actual heat loss} = A_W U_W + A_D U_D + A_R U_R$$

Combining these equations, the test for whether or not the areas satisfy the requirement is:

- For metal frames: $A_W U_W + A_D U_D + A_R U_R \leq 0.55A_F$
- For all other frames: $A_W U_W + A_D U_D + A_R U_R \leq 0.50A_F$

Examples of how the adjustment may be made are given in Chapter 7. However, within the elemental method, these formulae *cannot* be used to *increase* the areas above the standard area provision. If this is desired, a different approach must be used, such as the target U-value or the carbon index method.

2.3.2.2   The elemental method applied to extensions to dwellings

Unlike other methods, the elemental method can be applied to an extension to a dwelling. If the target U-value or the carbon index method is used, the extended dwelling must be considered as a whole, and the method applied to the complete newly enlarged structure. However, when applied to extensions, the elemental method does allow some flexibility. The ways in which the requirements can be satisfied are:

- Adhere to the maximum U-values given in Table 2.1 for all elements in the extension, including the area-weighted average U-value of the windows, doors and rooflights. Alternatively, use U-values which would give the same total rate of heat loss from the extension if the U-values in Table 2.1 had been used.
- *In addition*, ensure that *either*:
  (1)   the area of openings (i.e. windows plus doors plus rooflights) in the extension does not exceed the area of any original windows or doors which no longer exist or are no longer exposed because of the extension, plus 25% of the floor area of the extension, *or*
  (2)   the total area of openings in the new enlarged dwelling does not exceed the total area of openings in the existing dwelling prior to the extension, *or*
  (3)   the area of openings in the enlarged dwelling does not exceed 25% of the total floor area of the enlarged dwelling.

In order to interpret condition (1) above, first take the area of the original windows and/or doors which no longer exist or are no longer exposed because of the extension. Then take 25% of the floor area of the extension. Add these two areas together. The resulting total is the maximum permissible area of openings in the extension.

A further relaxation is allowed in the case of a small extension where the additional heated space has a maximum floor area of about $6\,m^2$. In such cases, the construction details of the extension will be acceptable if they have an energy performance as least as good as those of the existing dwelling. This relaxation would typically apply to a porch or small kitchen extension.

2.3.3 The target U-value method for dwellings

In order to demonstrate compliance using the target U-value method, it is necessary to calculate and compare two U-values. The first, U_T, is a target U-value. This is a theoretical index which is calculated from an initial basic formula, and which is then adjusted to allow for any design features that affect the energy consumption of the dwelling. The second, U_{AVG}, is the area-weighted average U-value of the actual dwelling, taking into account all exposed surfaces including walls, floors, roofs, windows, doors, rooflights and all elements adjacent to unheated spaces. The two values are then compared. The target U-value procedure can be arranged in several stages:

- Calculate an initial target U-value, U_1, using the standard formula
- Adjust U_1 for the efficiency of the proposed boiler, and obtain U_2
- For dwellings with metal window frames, allow for additional solar heat gain by adjusting U_2 to U_3
- For dwellings with sufficient south facing openings, allow for additional solar heat gains by adjusting U_3 to the target U-value, U_T
- Calculate the average U-value, U_{AVG}, of the proposed dwelling, using the actual U-values and areas of the elements.

The target U-value, U_T, is the overall maximum U-value of the dwelling, and as such, replaces the maximum U-values specified in Table 2.1 for the elemental method. Provided the average U-value, U_{AVG}, is less than (or at worst equal to) U_T, the requirement for compliance is satisfied.

The calculation procedure and the relevant formulae may be arranged as follows.

(1) Find U_1, the initial target U-value, from:

$$U_1 = \left[0.35 - 0.19 \frac{A_R}{A_T} - 0.10 \frac{A_{GF}}{A_T} + 0.413 \frac{A_F}{A_T} \right]$$

where U_1 is the initial target U-value
A_R is the area of exposed roof

A_{GF} is the ground floor area
A_F is the total floor area, including all storeys
A_T is the total area of all exposed elements.

(2)   Adjust U_1 according to the boiler efficiency ratio, f_e, found from:

$$f_e = \frac{\text{Proposed boiler SEDBUK}(\%)}{\text{Reference boiler SEDBUK}(\%)}$$

where the reference boiler SEDBUK is the minimum value as given in Table 2.2. A convenient source of SEDBUK values may be found on the internet web site www.SEDBUK.com. The adjusted target U-value, U_2, is then:

$$U_2 = F_e \, U_1$$

If there is no suitable data for the boiler, either because there is no SEDBUK, no seasonal efficiency value from SAP 2001, or if the heating system for the dwelling is not known, then the target U-value must be made more demanding by dividing by 1.15 to compensate for a possible higher carbon emission rate. In such cases, the adjusted target U-value, U_2, is obtained from:

$$U_2 = \frac{U_1}{1.15}$$

(3)   Allow for additional solar gain when the dwelling has metal windows. This is because the framing ratio (the fraction of the area of the window opening taken up by the frame material) is usually less for metal frames than for other frame materials. In the elemental method the allowance is given by means of the slightly higher maximum U-value for metal frames. In the target U-value method, if the windows have metal frames (including thermally broken frames), the extra solar energy is included by multiplying U_2 by a factor of 1.03 to obtain U_3. Thus:

For metal window frames: $U_3 = 1.03 \, U_2$
For all other frames: $U_3 = U_2$

(4)   Allow for additional solar gain when the area of glazed openings on the southern elevation exceeds that on the northern elevation. In this context, the area of the glazed openings includes the frames, and the definitions of the two elevations are:
Southern elevation: facing within an arc which is $30°$ either side of South
Northern elevation: facing within an arc which is $30°$ either side of North.
The solar adjustment factor, ΔS, is found from:

$$\Delta S = 0.04 \left[\frac{A_S - A_N}{A_{TG}} \right]$$

where A_S is the area of glazing facing south
A_N is the area of glazing facing north
A_{TG} is the total area of all glazed openings in the dwelling.

If A_S is less than A_N then ΔS is taken as zero.

(5) Convert U_3 to the final target U-value, U_T, using:

$$U_T = U_3 + \Delta S$$

(6) Find U_{AVG} from:

$$U_{AVG} = \frac{\Sigma AU}{A_T}$$

where ΣAU is the summation of area times U-value for all exposed elements
A_T is the total area of all exposed elements.

Finally, U_{AVG} and U_T are compared. If U_{AVG} is less than or equal to U_T then the requirement is satisfied, that is:

$$U_{AVG} \leq U_T$$

If the calculation results in a fail, i.e. U_{AVG} is greater than U_T, then the situation can be rectified either by taking action to reduce the U-values of exposed elements, or by reducing the area of openings. It may also be possible to increase U_T by selecting a boiler with a higher efficiency, or by increasing the proportion of glazing on the southern elevation. Care must be taken to repeat the calculation of U_T from the beginning, otherwise it is possible to introduce an error in the allowances for solar gain. Examples of the target U-value method are given in Chapter 8.

2.3.4 The carbon index method for dwellings

There are two indices used for measuring the amount of carbon dioxide put into the atmosphere by a dwelling. The *carbon factor* (CF) is the carbon dioxide emission in kilograms per year per m^2 of floor area. As this measures the actual quantity of carbon emitted, it follows that *decreasing* its value is good. The *carbon index* (CI) is calculated from the carbon factor, converting it from a linear to a logarithmic scale and also inverting it. The conversion formula was chosen to provide a more convenient scale for measuring the energy efficiency of a dwelling, in which *increasing* values represent higher energy efficiency and are therefore good. The formulae which connect these indices are:

$$CF = \frac{CO_2}{A_F + 45.0}$$

$$CI = 17.7 - 9.0\log_{10}(CF)$$

where CO_2 is the carbon dioxide emission rate in kilograms per year
A_F is the total floor area.

The procedure for determining the carbon index is:

- Follow the standard assessment procedure for the energy rating of dwellings, SAP 2001 edition [19], using the worksheets provided, to obtain the SAP rating of the dwelling
- Continue following the worksheet calculation to obtain the carbon factor and the carbon index.

The requirement is satisfied if:

$$CI \geq 8.0$$

As with the target U-value method, there is no restriction on the area or U-value of any individual element in the dwelling, thus providing substantial flexibility in design. Nevertheless the same considerations apply with respect to the provision of daylight and to the risk of cold internal surfaces and surface condensation. Thus, even if the minimum carbon index requirement of 8.0 is met, the constraints on the area of glazing and the poorest acceptable U-values given below still apply.

2.3.5   Constraints applicable to all three methods of demonstrating compliance

2.3.5.1   *Area of glazing*

All three methods of demonstrating compliance allow the area of windows, doors and rooflights to be reduced. However, consideration must also be given to the need to provide adequate daylight. As a general rule, the minimum standard for daylight provision occurs when the total area of openings is 17% of the total floor area. Anything less than this would be inadequate. Further information is available in BS 8206: Part 2 [20].

2.3.5.2   *Poorest acceptable U-values*

The flexibility built in to the target U-value and the carbon index methods allows the U values of some parts of the roof, walls or floor of a dwelling to be worse (i.e. higher) than the values given in Table 2.1, provided that this poorer performance is compensated for elsewhere. However, if the U-value of any part of a roof, wall or floor is too high, there is an increased risk of both surface and interstitial condensation. Consequently, regardless of the results of the target U-value or carbon index calculations, AD L1 gives values for poorest (i.e. maximum) acceptable U-values, as shown in Table 2.3.

Table 2.3 Poorest (i.e. maximum) acceptable U-values.

Element	Maximum acceptable U-value, W/m^2K
Part of roof	0.35
Part of exposed wall or floor	0.70

It should be noted that the values given in Table 2.3 would normally only be applied to parts of a roof, wall or floor. If they were applied to the whole of an element, it is very unlikely that compliance would be achieved, whatever compensatory measures had been taken elsewhere.

2.3.6   Additional requirements for fabric thermal performance

As overall standards of thermal insulation improve, so heat losses via thermal bridges and unwanted air leakage become more significant and therefore more important. Consequently there is a requirement to take action to minimise these effects.

2.3.6.1   *Thermal bridging*

Thermal bridges are most often due to:

● Gaps in insulation layers within the fabric
● Structural elements, especially lintels and frames (in timber, concrete, etc.)
● Joints between elements
● Joints around windows and doors.

Because of this, there is a requirement to construct in such a way as to avoid significant thermal bridges. This can be done in one of two ways:

● Demonstrating, by calculation, that the thermal performance is satisfactory; although there are several methods, none are simple (see, for example, [17], [18] and [55])
● Adopting robust construction details that have an authoritative recommendation, such as [2].

2.3.6.2   *Air leakage*

Air leakage is the unwanted passage of air through the building fabric, and must be distinguished from the planned ventilation which is necessary for health and safety. Air leakage is most often due to:

- Poor joints within or between elements of the external fabric
- The use of construction elements which are inherently porous to air movement
- Gaps around service pipes, ducts and flues where they penetrate the external fabric
- Ill-fitting windows, doors and rooflights.

The solution is to provide a continuous barrier to air movement around the habitable space (including separating walls and the edges of intermediate floors), penetrated only in those places where it is intentional. As with thermal bridges, this can be demonstrated by adopting published robust construction details. Alternatively, the dwelling can be pressure tested according to the procedure given in CIBSE TM 23 [21], the standard for acceptability being:

$$AP_{50} \leq 10 \ m^3/h/m^2$$

where AP_{50} is the air permeability in cubic metres per hour per square metre of external surface area at an applied pressure difference of 50 pascals.

2.3.7   Heating and hot water systems for dwellings

The requirements can be considered under four headings:

- Controls for space heating
- The provision of hot water by means of systems which incorporate hot water storage
- Insulation of pipes and ducts
- Commissioning, operating and maintenance instructions.

2.3.7.1   *Controls for space heating*

The guidance given in AD L1 is most suitable for systems in which heat is distributed from a central heat source. For such systems, it is necessary to consider the inclusion of zone controls, timing controls and boiler interlock controls. Although there are a large number of possible system types, Table 2.4 gives guidance for the most usual situations. No guidance is given on stand-alone heaters, whether supplied by solid fuel, gas or electricity.

Table 2.4   Controls for space heating.

Temperature zone controls	Timing zone controls	Boiler interlock controls
Control temperatures independently in areas with different heating needs. such as separate living and sleeping areas.	In most dwellings with two temperature zones, one timing zone is sufficient to control both temperature zones.	For gas and oil fired hot water central heating systems, the boiler must switch off when no heat is required, regardless of the type and location of thermostats.
For large dwellings, ensure that no temperature zone exceeds 150m² floor area. If so, subdivide into smaller zones.	For large dwellings, ensure that no timing zone exceeds 150m² floor area. If so, subdivide into smaller zones.	In systems controlled by thermostats, the boiler must only switch on when the thermostats call for space heating and/or hot water and/or heat for a storage vessel.
For small dwellings such as single-storey open-plan flats and bed-sitters, one temperature zone may be sufficient.	For small dwellings such as single-storey open-plan flats and bed-sitters, one timing zone may be sufficient.	Where thermostatic radiator valves are used, a room thermostat must also be provided to switch off the boiler when there is no demand for heat or hot water.
Provide temperature control by room thermostats *and/or* thermostatic radiator valves *or* any suitable device for sensing temperature. Temperature sensors must be linked to suitable control elements.	Provide separate timing control for space heating and water heating (except for combination boilers and solid fuel appliances).	

2.3.7.2   *The provision of hot water systems in dwellings*

Although there are several acceptable ways of providing hot water, AD L1 provides guidance only for systems which include integral or separate hot water storage. Possible ways of satisfying the requirement, together with the relevant standards, are given in Table 2.5.

Table 2.5 Standards for hot water systems.

Meet the insulation standards of the most appropriate of these listed standards	BS 1566 [22] BS 699 [23] BS 3198 [24] BS 7206 [25]
or	
In ordinary cases, use insulating vessels with a factory-applied coating of polyurethane foam to meet this stated specification	Minimum thickness of insulation, 35mm Maximum density of insulation, 30kg/m^3
and	
Avoid excessive boiler firing and primary circuit losses, and enable efficient operation. For indirectly heated hot water storage systems, size the heat exchanger according to the most appropriate of these standards, and feed it by a pumped primary system	BS 1566 [22] BS 3198 [24] BS 7206 [25]
or	
For primary storage systems meet the requirements of the WMA standard for thermal stores.	Performance specification for thermal stores Waterheater Manufacturers Association, 1999 [26]

2.3.7.3   *Alternative approach for space heating and the controls for hot water systems*

Good practice guide GPG 302 [27] and BS 5864 [28] are both relevant. If an installation follows the recommendations of either of these documents, and also includes the zoning, timing and interlock features described above, then it should meet the requirement.

2.3.7.4   *Commissioning, operating and maintenance instructions for heating and hot water systems*

When completed, heating and hot water systems should be inspected and commissioned to ensure that they:

- Operate correctly and to specification
- Comply with health and safety requirements
- Operate efficiently with respect to the need to conserve fuel and power
- Achieve compliance with the requirements of AD L1.

In this context, commissioning means moving from static completion to full operation, and includes:

- Setting the system to work
- Regulation (i.e. repetitive testing and adjustment) to achieve correct performance
- Calibration, setting up and testing of associated automatic control systems
- Recording of the system settings and performance test results that have been accepted as satisfactory.

Responsibility for achieving compliance with AD L1 lies with the person carrying out the work. Such a person may be a:

- Developer or contractor who has directly carried out the work
- Sub-contractor engaged by a developer or contractor
- Specialist organisation directly engaged by a private client.

Written certification that commissioning has been successfully carried out and that compliance with Part L1 (b) and (d) has been achieved must be provided and made available to both the client and the building control body. The person providing the certificate should normally have a recognised qualification, and should be the person responsible for achieving compliance, i.e. the developer, contractor, sub-contractor or specialist. If the person providing the certificate does not have a relevant qualification, or if a suitably qualified certifier is not available, a written declaration of successful commissioning must still be obtained and made available to both the client and the building control body.

Information on the operation and maintenance of the heating and hot water services must be provided and given to the building owner and/or occupier. Operating and maintenance instructions must:

- Be provided for each new dwelling
- Be provided for an existing dwelling when the systems are substantially altered
- Be in an accessible format
- Be directly related to the system or systems in the dwelling
- Explain to householders how to operate the systems, and what routine maintenance is advisable, so that the systems perform efficiently in terms of the conservation of fuel and power.

2.3.7.5 *Insulation of pipes and ducts*

The insulation of pipes and ducts is required in order to conserve heat and thus maintain the temperature of water or air being supplied to the heat emitters of a heating system. Insulation in hot water services systems is also required to prevent excessive losses between useful draw-off points. Guidance on methods of achieving the requirements is given in Table 2.6.

Table 2.6   Insulation of pipes and ducts.

Type of pipe or duct	Insulation requirement
Space heating pipe-work located outside the insulation layer(s) of the building fabric	Wrap the pipe with insulation which has maximum thermal conductivity of 0.035W/mK at 40°C, to a minimum thickness equal to the outside diameter of the pipe or duct, up to a maximum thickness of 40 mm
Hot pipes connected to hot water storage vessels, including the vent pipe and the primary flow and return to the heat exchanger	As above for space heating pipe-work, for at least 1 metre from the point of connection to the storage vessel (or to the point where the pipes become concealed)
Pipes and warm air ducts	Provide insulation in accordance with the recommendations of BS 5422 [29]

2.3.8   Lighting systems for dwellings

There is a requirement to apply energy conservation principles to both internal and external lighting by providing systems with appropriate lamps and sufficient controls. However, although the regulation itself appears to demand that *all* lamps are 'energy efficient', the guidance allows some flexibility. Furthermore, the requirement for sufficient controls does *not* apply to internal lighting, only to external lighting.

   Care must be taken to interpret the term 'luminous efficacy' correctly. When referring to dwellings in AD L1, it refers to the amount of light given out by the lamp. The definition, therefore, is:

$$\text{Luminous efficacy} = \frac{\text{Total lumens emitted by lamp}}{\text{Total circuit-watts consumed by lamp}}$$

Note that no account is taken of the fitting or lampshade in which the lamp is placed.

2.3.8.1   Internal lighting

The guidance recommends the following approach:

   At a minimum number of locations, provide fixed lighting comprising
   - *either*   basic lighting outlets
   - *or*   complete luminaires
   that will *only* take lamps of luminous efficacy greater than 40 lumens per circuit-watt.

The minimum number of locations must include those that are expected to have most use, and can be calculated from the number of rooms in the dwelling by the formula:

$$\text{Minimum number of locations} = \frac{\text{Number of rooms}}{3}$$

the result being rounded up to the nearest whole number. Thus, for three rooms the minimum number of locations is one, whereas for four rooms the calculation gives $1\frac{1}{3}$, which is rounded up to two. When counting the rooms in the house, note that:

- Hall, stairs and landing count as one room, even if they contain more than one light fitting
- An integral conservatory in a new dwelling is included in the count, but in other cases a conservatory is excluded
- Garages lofts and outhouses are excluded.

Note also that in the definition of luminous efficacy, circuit-watt includes the electrical power consumed by the lamps plus all associated control gear and power factor equipment. Typically, it may be expected that the 40 lumens per circuit-watt minimum would be achieved by fluorescent tubes and compact fluorescent lamps, *but not* by GLS tungsten lamps with bayonet caps or Edison screw bases.

Further guidance on internal lighting is given in General Information Leaflet GIL 20, *Low energy domestic lighting* [30].

2.3.8.2 *External lighting*

The requirement is in respect of systems fixed to the building, and includes porches but not lighting in garages and car ports. The guidance suggests two alternative solutions:

(1) Provide controls which automatically extinguish the system when there is enough daylight *and* when the the system is not required at night, *or*
(2) Install sockets that will *only* take lamps of luminous efficacy greater than 40 lumens per circuit-watt. As for internal lighting, this means that typically fluorescent tubes and compact fluorescent lamps (*but not* GLS tungsten lamps with bayonet caps or Edison screw bases) will be acceptable.

2.3.9 Conservatories

A conservatory is defined as a space that has not less than three-quarters of the area of its roof and not less than one-half of the area of its external walls made of translucent material. It is also necessary to define 'separation'. In this context, separation between a dwelling and a conservatory means:

- Separating walls and floors insulated to at least the same degree as the exposed walls and floors of the dwelling

- Separating windows and doors with the same U-value and draught-stripping provisions as the exposed windows and doors elsewhere in the dwelling.

A conservatory may be attached to and constructed as part of a new dwelling, or it may be attached to an existing dwelling. The requirements of these two cases differ slightly.

2.3.9.1   A conservatory attached to and built as part of a new dwelling

There are three possibilities, one where there is no separation between the conservatory and the dwelling, and the other two where there is separation:

- No separation – the conservatory should be treated as an integral part of the dwelling
- Separation, unheated – energy savings can be achieved if the conservatory is not heated
- Separation, heated – if fixed heating installations are proposed, they should have their own separate temperature and on/off controls.

2.3.9.2   A conservatory attached to an existing dwelling

Reasonable provision must be made to limit heat loss from the dwelling. Ways of doing this depend on whether or not the opening to the conservatory is newly created or enlarged:

- If the opening is *not* to be enlarged, retain the existing separation
- If the opening *is* to be newly created or enlarged, provide separation the same as or equivalent to windows and doors having the same maximum average U-value given in Table 2.1.

2.4   AD L1 – Section 2 Work on Existing Dwellings

As well as extensions and conservatories, the regulations also cover certain other categories of work on existing dwellings. There are four such categories:

- Replacement of controlled services or fittings
- Material alterations
- Material change of use
- Historic buildings.

2.4.1   Replacement of controlled services or fittings

A controlled service or fitting is defined in Regulation 2(1) of the Building Regulations (2000 as amended 2001) as:

'...a service or fitting in relation to which Part G, H, J or L of Schedule 1 imposes a requirement'.

In this context, whether or not replacement of controlled services or fittings falls within the definition of building work (and is therefore subject to the requirements of Part L) is stated in Regulation 3(1), as qualified in Regulation 3(1A), as follows:

'The provision or extension of a controlled service or fitting
   (a)   in or in connection with an existing dwelling, *and*
   (b)   being a service or fitting in relation to which Part L1, but not Parts G, H, or J, of Schedule 1 imposes a requirement shall only be building work where that work consists of the provision of any of the following':

- Window
- Rooflight
- Roof window
- Door, glazed to more than 50% of its total area, including its frame, measured internally
- Space heating or hot water service boiler
- Hot water vessel.

Part L1 applies to replacement work on controlled services or fittings when:

- Replacing old with new identical equipment
- Replacing old with new but different equipment
- The work is solely in connection with controlled services or includes work on them.

Ways of satisfying the requirements of Part L1 may depend on the circumstances of the particular case. Specific guidance is as follows.

2.4.1.1   Windows, doors and rooflights

The requirement does *not* apply to repair work on *parts* of these elements such as:

- Replacing broken glass
- Replacing sealed double-glazing units
- Replacing rotten framing members.

However, where these elements are to be replaced rather than repaired, it becomes necessary to provide new draught-proofed ones with either:

- An average U-value not exceeding the appropriate value in Table 2.1, *or*
- A centre-pane U-value not exceeding 1.2 W/m^{2}K.

The replacement work should comply with the requirements of Parts L and N. Furthermore, after completion of the work, the building should not have a worse level of compliance with other applicable parts of Schedule 1, including Parts B, F and J.

2.4.1.2   Heating boilers

If the dwelling has a floor area greater than $50 \, m^2$, the new boiler should satisfy the same requirements as for a new dwelling. This includes:

- For ordinary gas or oil boilers, adherence to the minimum SEDBUK ratings given in Table 2.2
- For back boilers, adherence to a minimum SEDBUK rating that is 3 percentage points less than the appropriate value in Table 2.2
- For solid fuel boilers, provision of a boiler having an efficiency not less than that recommended for its type in the HETAS [31] certification scheme.

2.4.1.3   Hot water vessels

When replacing hot water vessels, new equipment should be provided which would satisfy the requirements for a new dwelling.

2.4.1.4   Boiler and hot water controls

To ensure that replacement boilers (except solid fuel boilers) and hot water vessels achieve a reasonable and acceptable seasonal efficiency, it may be necessary to replace or provide:

- The time switch or programmer
- The room thermostat
- The hot water vessel thermostat
- A boiler interlock
- Fully pumped circulation.

Advice and information on how this can be done is given in Good Practice Guide GPG 302, section 3 [27].

2.4.1.5   Alternative approach using the carbon index

As an alternative to the procedures of the above four paragraphs, it may be acceptable to follow the guidance in Good Practice Guide GPG 155 [32], provided that an equivalent improvement in the carbon index of the dwelling is achieved.

2.4.1.6 *Commissioning, and operating and maintenance instructions*

If heating and/or hot water systems are altered or replaced, it is reasonable to expect that commissioning would be carried out, and operating and maintenance instructions provided, as if for a new dwelling.

2.4.2 Material alterations

It is necessary to define the terms 'Material alteration' and 'Relevant requirement'.

Material alteration (Regulation 3(2))
   'An alteration is material for the purposes of these regulations if the work, or any part of it, would at any stage result –
   (a) in a building or controlled service or fitting not complying with a relevant requirement where previously it did, or
   (b) in a building or controlled service or fitting which before the work commenced did not comply with a relevant requirement, being more unsatisfactory in relation to such a requirement.'

Relevant requirement (Regulation 3(3))
As used in Regulation 3(2):

   '... relevant requirement means any of the following applicable requirements of Schedule 1, namely –
   Part A (structure)
   Paragraph B1 (means of warning and escape)
   Paragraph B3 (internal fire spread – structure)
   Paragraph B4 (external fire spread)
   Paragraph B5 (access and facilities for fire service)
   Part M (access and facilities for disabled people).'

The consequence of these definitions is that an alteration is only considered material if it affects the existing building with respect to any one or more of Parts A, B1, B3, B4, B5 or M. However, as soon as it is established that an alteration is material, then account should be taken of:

- All the relevant requirements of Schedule 1, including Parts L1, F and J
- Insulation of roofs, floors and walls
- Sealing measures
- Controlled services and fittings (as above).

Ways of satisfying the requirements of Part L1 may depend on the circumstances of the particular case. Specific guidance is as follows.

2.4.2.1 Roof insulation

If the material alteration includes substantial replacement of any of the major elements of a roof structure, insulate to the U-value standard of a new dwelling.

2.4.2.2 Floor insulation

If the structure of a ground floor or exposed floor is to be substantially replaced or re-boarded, and if the room is heated, insulate to the U-value standard of a new dwelling.

2.4.2.3 Wall insulation

Provide a reasonable thickness of insulation when substantially replacing:

- Complete exposed walls
- The external rendering or cladding of an exposed wall
- The internal surface finishes of an exposed wall
- The internal surfaces of separating walls.

2.4.2.4 Sealing measures

When carrying out any of the above work on roofs, floors or walls, include reasonable sealing measures to improve airtightness, but remember to take into account the requirements of Parts F and J.

2.4.2.5 Controlled services or fittings

These are dealt with above.

2.4.3 Material changes of use

A material change of use is defined differently from a material alteration. According to Regulation 5:

> '...for the purposes of these Regulations, there is a material change of use where there is a change in the purpose for which or the circumstances in which a building is used, so that after the change –
> (a) the building is used as a dwelling, where previously it was not;
> (b) the building contains a flat, where previously it did not;
> (c) the building is used as an hotel or a boarding house, where previously it was not;
> (d) the building is used as an institution, where previously it was not;
> (e) the building is used as a public building, where previously it was not;
> (f) the building is not a building described in Classes I to VI in Schedule 2, where previously it was;

(g)   the building, which contains at least one dwelling, contains a greater or lesser number of dwellings than it did previously.'

Regulation 6 includes a list of all those parts of Schedule 1 that apply when works comprising a change of use are undertaken. The list includes Part L1. It also includes two other parts of relevance to the conservation of fuel and power and to which particular attention should also be paid: Part F (ventilation) and Part J (combustion appliances). As far as Part L1 is concerned, when undertaking material alterations, account should be taken of:

- Accessible lofts
- Insulation of roofs, floors and walls
- Sealing measures
- Lighting
- Controlled services and fittings (as above).

Clearly, if the whole of a building is subject to a material change of use, then the whole building should comply with all the relevant parts of Schedule 1 that are listed in Regulation 6. If the change of use applies to only part of a building, then in general only that part must comply. Ways of satisfying the requirements of Part L1 may depend on the circumstances of the particular case, and are mostly the same as for a material alteration. Specific guidance is as follows.

2.4.3.1   Accessible lofts

Where the existing insulation in accessible lofts is worse than $0.35 \, W/m^2 K$, replace or add extra insulation to upgrade the U-value to a maximum of $0.25 \, W/m^2 K$.

2.4.3.2   Roof insulation

If the material alteration includes substantial replacement of any of the major elements of a roof structure, insulate to the U-value standard of a new dwelling.

2.4.3.3   Floor insulation

If the structure of a ground floor or exposed floor is to be substantially replaced or re-boarded, and if the room is heated, insulate to the U-value standard of a new dwelling.

2.4.3.4   Wall insulation

Provide a reasonable thickness of insulation when substantially replacing:

- Complete exposed walls
- The external rendering or cladding of an exposed wall
- The internal surface finishes of an exposed wall
- The internal surfaces of separating walls.

2.4.3.5 Sealing measures

When carrying out any of the above work on roofs, floors or walls, include reasonable sealing measures to improve airtightness, but remember to take into account the requirements of Parts F and J.

2.4.3.6 Lighting

Provide lighting in accordance with requirements for a new dwelling.

2.4.3.7 Controlled services or fittings

These are dealt with above.

2.4.4 Historic buildings

Historic buildings include:

- Listed buildings
- Buildings situated in conservation areas
- Buildings of architectural and historical interest and which are referred to as a material consideration in a local authority development plan
- Buildings of architectural and historical interest within national parks, areas of outstanding natural beauty, and world heritage sites.

Any work on an historic building must balance the need to improve energy efficiency against the following factors:

- The need to avoid prejudicing the character of the historic building
- The danger of increasing the risk of long-term deterioration of the building fabric
- The danger of increasing the risk of long-term deterioration of the building's fittings
- The extent to which energy conservation measures are a practical possibility.

Advice on achieving the correct balance should be sought from the conservation officer of the local authority. Advice from other published sources, e.g. PPG15 [33], BS 7913 [34] and SPAB Information sheet 4 [35], may also be appropriate, particularly regarding:

- Restoration of the historic character of a building that had been the subject of inappropriate alteration, such as the replacement of windows, doors or rooflights
- Rebuilding of a former historic building, which may have been damaged or destroyed due to some mishap (such as a fire), or in-filling a gap in a terrace
- Providing a means for the fabric of an historic building to 'breathe' so that moisture movement may be controlled and the potential for long-term decay problems reduced.

3 The Conservation of Fuel and Power in Buildings other than Dwellings

Part L2 of Schedule 1 to the 2001 Regulations is concerned with the conservation of fuel and power in buildings other then dwellings. This is supported by Approved Document L2 which includes the following topics:

- Compliance by the elemental method
- Compliance by the whole-building method
- Compliance by the carbon emissions calculation method
- Limiting thermal bridging and air leakage
- Avoiding solar overheating
- Heating systems
- Controls for space heating and hot water systems
- Insulation of pipes, ducts and vessels
- Inspection and commissioning of the building services systems
- Lighting
- Air conditioning and mechanical ventilation (ACMV).

The purpose of Part L2 is to minimise the environmental impact of buildings with respect to the depletion of fuel reserves and the degradation of the atmosphere by carbon dioxide emissions.

3.1 The legal requirement for the conservation of fuel and power in buildings other than dwellings

The legal regulation requires that a building (other than a dwelling) should be so designed and constructed as to make reasonable provision for the conservation of fuel and power, and that this shall be achieved by attention to all of the following eight points:

- Limiting heat losses and gains through the fabric of the building
- Limiting the heat loss from hot water pipes and hot air ducts used for space heating, and limiting the heat loss from hot water vessels and hot water service pipes
- Ensuring that the space heating and hot water systems which are provided are energy efficient

- Limiting exposure to solar overheating
- Where a floor area greater than $200\,\text{m}^2$ is served by an ACMV system, ensuring that no more energy needs to be used than is reasonable in the circumstances
- Limiting the heat gains to chilled water and refrigerant vessels and pipes and air ducts that serve air conditioning systems
- Providing lighting systems which use energy efficiently
- Ensuring that, by the provision of sufficient information with the relevant services, the building can be operated and maintained in such a manner as to use no more energy than is reasonable in the circumstances.

Approved Document L2 gives general and specific guidance on the energy efficient measures which will satisfy the requirements.

3.2 AD L2 – Section 0 General Guidance

The general guidance for Part L2 is broadly similar to that given for Part L1. However, because of the large range of building types in the non-domestic sector, the guidance is broader in scope, and there are some differences in the detail.

Heat losses through walls, roofs, floors, windows and doors, etc. must be kept within acceptable limits by means of suitable levels of thermal insulation. Also, where it is possible and appropriate, any beneficial gains from solar heat and more efficient heating systems should be taken into account. However, it is also expected that, where necessary, provision be made to limit heat gains in summer.

Heat may also be lost by unplanned air leakage through the fabric of a building, particularly around doors, windows and junctions between elements or components. This requires attention not only to draught stripping of doors and windows, but also consideration of potential air leakage paths through all other elements of the building.

Heat losses from hot water pipes, hot air ducts, hot water vessels, and all their connections must also be controlled by the application of sufficient insulation. If, however, a particular component contributes to the space heating in an efficient manner, the insulation may be omitted.

The energy efficiency of the systems which provide space heating and hot water has two aspects. Firstly, the equipment itself must generate heat from fuel in an efficient manner, and secondly the equipment must have controls for space and water temperatures and for the timing of its operation. Furthermore, the systems must have been appropriately commissioned and be capable of being operated in an efficient manner by the user.

In order to limit solar overheating, a combination of techniques may be used. These include passive measures such as limiting the area of unshaded glazing and designing the external fabric so that it limits and delays heat penetration, plus active measures such as night ventilation.

Conservation of energy in mechanically ventilated and air conditioned buildings requires that the energy demand is limited and that the energy which is needed is supplied efficiently. The energy demand includes heating, cooling, air circulation, water and refrigerants. The energy supply must include efficient plant, controls for timing, temperature and flow, and energy consumption metering. All services must have been appropriately commissioned.

Heat gains to chilled water and refrigerant vessels, and to the ducts and pipes which may be carrying chilled air or fluids, must be limited by suitable thicknesses of insulation, with vapour barriers to prevent the insulation becoming damp.

There are four requirements for lighting systems:

(1) Systems must use energy efficient lamps and luminaires
(2) The switching must be manual or automatic, or a combination of manual and automatic
(3) The lighting systems must have their own energy consumption metering
(4) The complete lighting system must have been appropriately commissioned.

The energy efficiency of heating and hot water systems can be affected by the manner in which they are operated. Incorrect usage may result in poor energy performance from an otherwise energy efficient system. Consequently, there is a requirement to provide easily understandable information that will enable the occupier to operate and maintain the building and its building services so that they do not waste energy. This information should also include the results of performance tests.

3.2.1   Carbon and carbon dioxide indices

In order to be consistent with the general objective of Part L, which is to reduce pollution of the atmosphere (especially by carbon dioxide) caused by energy consumption in buildings, the fuel consumed by a building is most conveniently expressed in terms of the amount of carbon that has been generated. In AD L2, therefore, performance targets are expressed in kilograms of carbon, and not in terms of an energy unit such as gigajoules or megawatt-hours. Occasionally, there may be a preference for expressing performance in terms of gaseous carbon dioxide rather than solid carbon. The conversion factor from one to the other is the ratio of the molecular weights, i.e. 44/12. Thus, 6 kg per square metre of carbon per year is equivalent to $6 \times (44/12) = 22$ kg of carbon dioxide per square metre per year.

3.2.2   Special cases

3.2.2.1   Exemptions

There are two types of exemption from the requirements of L2: when there are low levels of heating and/or when there are low levels of use.

3.2.2.1.1   Low levels of heating

A building with a heat requirement which does not exceed $25\,W/m^2$ may be regarded as having a low heat requirement, and therefore does not require measures to limit heat loss through the fabric. In such a case, the fabric insulation is likely to be chosen for operational reasons. Typical examples are:

- A warehouse where general heat is provided as a protection against frost or condensation damage to goods, and where higher temperatures are only necessary for local work stations
- A cold store, where the major concern may be to avoid heat gain rather than heat loss through the building fabric.

3.2.2.1.2   Low levels of use

A building which, because of its function, is used for a smaller number of hours than is normal, may where appropriate be fitted with heating and lighting systems to a lower standard than that required by AD L2. However, the building must adhere to the fabric insulation standards of AD L2. A typical example is:

- A building used solely for worship at set times.

3.2.2.2   Historic buildings

Historic buildings are considered in section 3.6.4.

3.2.2.3   Buildings constructed from sub-assemblies

A new building, however it is constructed, must normally comply with all the requirements of Schedule 1. However, with respect to the external fabric of buildings constructed from sub-assemblies, reasonable provisions for the conservation of fuel and power may vary according to the circumstances of the particular case. Examples where this may apply are:

- A building created from the external fabric sub-assemblies of an existing building by dismantling, transporting and re-erecting on the same site; this would normally be considered to meet the requirements
- A building which uses external fabric assemblies manufactured before 31 December 2001, or obtained from other premises; this would normally be considered to meet the requirements if the fabric thermal resistance, or the predicted annual energy consumption, satisfies the relevant requirements of Approved Document L, 1995 edition.

Enclosed heating and cooling links (which may also be made from sub-assemblies) should be insulated and made airtight to the same standards as the buildings themselves. In temporary accommodation, the requirements that must be satisfied by the heating and lighting may depend on the particular case. The normal expectation would be:

- *For heating and hot water systems* provide on/off time and temperature controls as specified in sections 3.3.2.8 and 3.3.2.9
- *For general and display lighting* follow the guidance in sections 3.3.2.11.2 and 3.3.2.11.3.

3.2.2.4   Mixed use development

Where part of a building is used as a dwelling and the rest of it is used for other purposes, the requirements for dwellings must also be taken into consideration.

3.3   AD L2 – Section 1 Design

The are a number of general considerations which should be applied to a building design project. These include:

- Preparing designs for the building and its services which are appropriate to the need to achieve energy efficiency
- Providing information so that the performance of the building in use may be assessed
- Making provisions in the design of the building services installations which facilitate inspection and commissioning.

Reference may be made to the CIBSE Guide on energy efficiency in buildings [36].

It is permissible and also perhaps preferable, if a building is large and complex, to consider the building in parts and apply the principles and requirements for energy conservation to each part. Also, where a building has alternative building services (e.g. dual fuel boilers, combined heat and power with standby boilers, etc.), the building should meet the requirements of AD L2 in all possible operating modes.

3.3.1   Specific guidance

For specific guidance, Approved Document L2 offers three methods for demonstrating that reasonable provision has been made for limiting heat loss through the building fabric:

The elemental method has the advantage that it does not require complex calculations. Each aspect of the building is considered individually, and a minimum level of performance must be achieved in each of the elements. Some flexibility is provided by allowing trade-off between different elements of the construction, and between insulation standards and heating system performance.

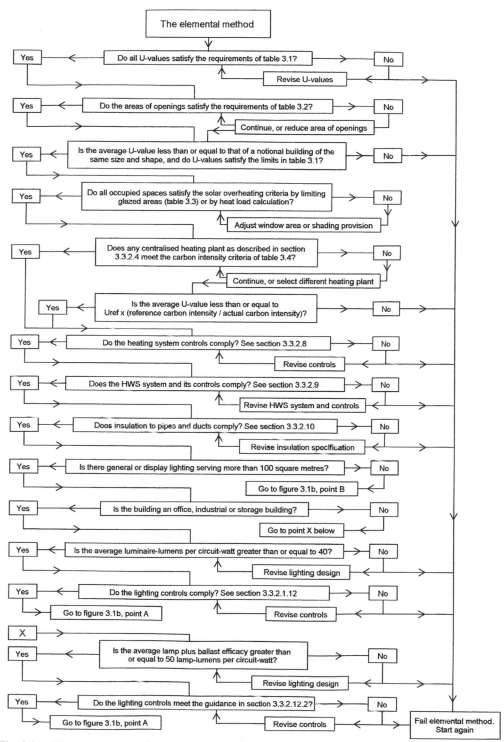

Fig. 3.1a Flowchart for the elemental method, Part L2.

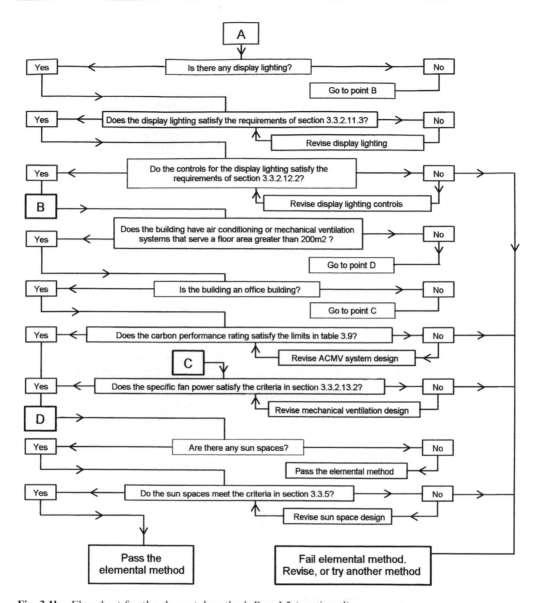

Fig. 3.1b   Flowchart for the elemental method, Part L2 (continued).

The whole-building method considers the performance of the whole building. It is applicable to office buildings, but similar alternative methods are available for schools (see DfEE Building Bulletin 87 [47]) and hospitals (see NHS Estates Guides [48]).

The carbon emissions calculation method considers the performance of the whole building and can be applied to any building type. To comply, the annual carbon emissions from the proposed building should be no greater than those from an equivalent notional building that satisfies the criteria of

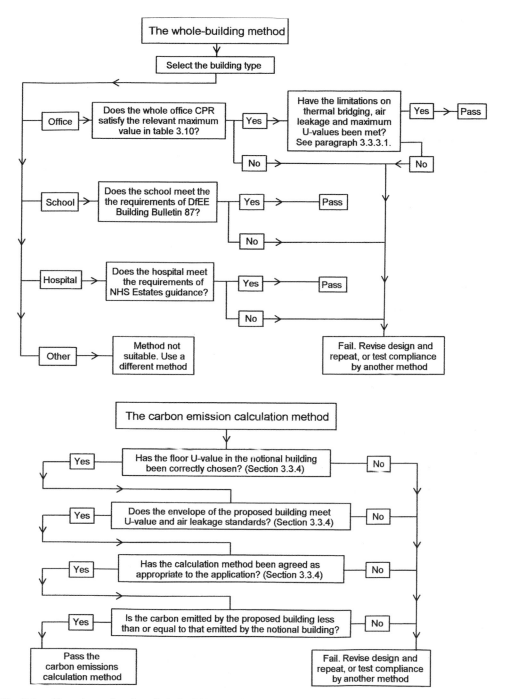

Fig. 3.1c Flowcharts for the whole-building method and the carbon emissions calculation method.

the elemental method. The carbon emissions from both the proposed building and the notional building must be estimated by means of an appropriate calculation tool.

Although there is no obligation to adopt any particular method or solution for satisfying the requirements, the Approved Document provides a check-list, in the form of a 'Summary Guide', that helps the designer choose the most suitable of the above three methods. This summary guide is shown in flow chart form in Figs 3.1a, 3.1b and 3.1c.

3.3.2   The elemental method for buildings other than dwellings

In the elemental method, compliance is demonstrated by specifying:

- Maximum U-values for walls, floors and roofs and rooflights
- Maximum area-weighted U-values for windows, personnel doors and rooflights
- Maximum combined area of windows and doors (as a percentage of exposed wall area)
- Maximum area of rooflights (as a percentage of roof area)
- Minimum performance standards for all energy consuming building services
- Design details to meet air leakage standards and to avoid solar overheating.

With regard to windows within part L2 the following should be noted:

- Display windows, shop entrance doors and similar glazing are exempt from the maximum U-value standard of Table 3.1, but not from the maximum area standards of Table 3.2
- The terms rooflight and roof window are synonymous
- For the purposes of calculating maximum areas, dormer windows in a roof may be included in the rooflight area.

Maximum U-values are shown in Table 3.1.

With the exception of vehicle access doors, Table 3.1 is the same as Table 2.1 for dwellings, and so the illustration of these U-values in Fig. 2.2 is still applicable. Also, the same comments apply with respect to unheated spaces and metal frame windows.

The area-weighted U-value of the windows, doors and rooflights can be calculated from the formula:

$$\text{Average U} - \text{value}, U_{AV} = \frac{A_W U_W + A_D U_D + A_R U_R}{A_W + A_D + A_R}$$

Table 3.1 Maximum U-values for the elemental method.

Exposed element	Maximum U-value, W/m²K	Comment
Pitched roof, insulation between rafters	0.20	Any part of a roof having a pitch of 70° or more can be considered as a wall
Pitched roof with integral insulation	0.25	
Pitched roof, insulation between joists	0.16	
Flat roof	0.25	Roof of pitch 10° or less
Walls, including basement walls	0.35	
Floors, including ground floors and basement floors	0.25	
Windows, personnel doors and roof windows, glazing in metal frames, area-weighted average for the whole building	2.20	A roof window may be considered as a rooflight. The higher U-value for metal frames allows for extra solar gain due to greater glazed proportion
Windows, personnel doors and roof windows, glazing in wood or PVC frames, area-weighted average for the whole building	2.00	
Rooflights	2.20	Applies only to the unit, excluding any upstand which must be insulated
Vehicle access and similar large doors	0.70	

where A_W, A_D and A_R are respectively the areas of windows, doors and rooflights

U_W, U_D and U_R are their U-values.

The maximum areas of windows, doors and rooflights depend on the building type, and are given in Table 3.2.

Table 3.2 Maximum areas of windows and rooflights.

Building type	Maximum window and door area as % of exposed wall area	Maximum rooflight area as % of area of roof
Residential buildings where people temporarily or permanently reside	30	20
Places of assembly, offices and shops	40	20
Industrial and storage buildings	15	20
Vehicle access doors and display windows and similar glazing	As required	

3.3.2.1   Adjustments allowed by the elemental method

The elemental method allows the U-values of the construction elements and the areas of windows, doors and rooflights to be varied from the values given in Tables 3.1 and 3.2, provided that the rate of heat loss from the proposed building is no worse than that from a notional building of the same size and shape that does meet the maximum U-values and maximum areas in these tables. There are, however, four constraints on the allowable variations:

(1)   If the U-value of any part of a roof, wall or floor in the proposed building is greater than the value given in Table 3.1, it must nevertheless not exceed the following absolute maxima:

- Any part of a roof: maximum acceptable U-value = 0.35 W/m^{2}K
- Any part of an exposed wall or floor: maximum acceptable U-value = 0.70 W/m^{2}K

(2)   If in the proposed building the floor has no added insulation and its U-value is better (i.e. lower) than the value given in Table 3.1, then this lower value must be used in the calculation for the notional building.

(3)   If the areas of any of the openings in the proposed building are less than the maxima in Table 3.2, then the average U-value of the roof, wall or floor must not exceed the relevant value in Table 3.1, by more than 0.02 W/m^{2}K.

(4)   No more than half of the allowable rooflight area can be used to increase the area of windows and doors. The increase is not calculated on a simple area basis, but by taking into account the difference in U-values between the roof and the wall. The method is given in Appendix E to AD L2, and is demonstrated in the example in Chapter 10.

3.3.2.2   Additional requirements for fabric thermal performance

As overall standards of thermal insulation improve, so heat losses via thermal bridges and unwanted air leakage become more significant and therefore more important. Consequently there is a requirement to take action to minimise these effects.

3.3.2.2.1   Thermal bridging
Thermal bridges are most often due to:

- Gaps in insulation layers within the fabric
- Structural elements, especially lintels and frames (in timber, concrete, etc.)
- Joints between elements
- Joints around windows and doors.

Because of this, there is a requirement to construct in such a way as to avoid significant thermal bridges. This can be done most easily by adopting robust construction details that have an authoritative recommendation. Although the detailing given in the 'robust construction details' document [2] is intended mainly for dwellings, it is also relevant to other buildings of similar construction and with similar internal environmental conditions. The detailing may not be satisfactory for a building with an unusual temperature or humidity regime. Alternatively, it may be possible to demonstrate by calculation that thermal bridging is not a problem and that the thermal performance is satisfactory. Although there are several methods of doing this, none are simple. The method described in IP 17/01 [55] is one of the least complex.

3.3.2.2.2   Air leakage
Air leakage is the unwanted passage of air through the building fabric, and must be distinguished from the planned ventilation that is necessary for health and safety. Air leakage is most often due to:

- Poor joints within or between elements of the external fabric
- The use of construction elements which are inherently porous to air movement
- Gaps around service pipes, ducts and flues where they penetrate the external fabric
- Ill-fitting windows, doors and rooflights and junctions around them.

These problems can be alleviated by providing sealing measures such as a continuous barrier to air movement in contact with the insulation layer over the whole thermal envelope, including separating walls and the edges of intermediate floors, and penetrated only in those places where it is intentional. For domestic type constructions, some suitable design details and installation practice are available in the robust details publication. For metal cladding and roofing systems, design guidance is given in the MCRMA Technical Report No. 14 [37]. The sealing of gaps around service penetrations and the draught-proofing of external doors and windows must also be attended to.

For buildings of any size, the rules for showing compliance with the requirements of AD L2 for resistance to air leakage are as follows:

- With effect from 1 October 2003, carry out an air leakage test in accordance with CIBSE TM 23 [21] to show that:

$$AP_{50} \leq 10 \text{ m}^3/\text{h}/\text{m}^2$$

where AP_{50} is the air permeability in cubic metres per hour per square metre of external surface area at an applied pressure difference of 50 pascals.
- In the period up to 30 September 2003, if the initial test results are unsatisfactory and appropriate remedial work has been carried out, show that:

○ *Either* there has been an improvement of at least 75% of the difference between the initial test result and the target standard of 10 m^3/h/m^2, i.e.

$$\text{Improvement} = 0.75 \, (\text{Initial test result} - 10) \, \text{m}^3/\text{h/m}^2$$

○ *Or* the re-test gives a result of

$$\text{AP}_{50} \leq 11.5 \, \text{m}^3/\text{h/m}^2$$

For buildings of less than 1000 m^2 gross floor area, the rules for showing compliance with the requirements of AD L2 for resistance to air leakage are:

- *Either* carry out air leakage tests and satisfy the requirements as above for a building of any size
- *Or* provide certificates or declarations that the design details, construction techniques and the execution of the work are such that reasonable conformity with the air leakage requirement can be expected.

3.3.2.2.3 Avoidance of solar overheating

Solar overheating can manifest itself in two ways. In naturally ventilated spaces it can lead to uncomfortably or even unacceptably high internal temperatures. In spaces with mechanical ventilation or cooling it can lead to excessive cooling capacity, and if the cooling capacity is insufficient, to high internal temperatures as well. AD L2 requires that both these problems should be addressed. The general objectives are therefore that buildings should be constructed so that:

- For occupied and naturally ventilated spaces, there should be no overheating when they are subjected to a 'moderate level' of internal heat gain
- For occupied spaces with mechanical ventilation or cooling, there should be no need for excessive cooling plant capacity in order to maintain the desired conditions.

The methods by which these objectives may be achieved could include some or all of:

- Specification of an appropriate type and quantity of glazing
- The incorporation in the design of passive measures such as shading
- The use of night ventilation in conjunction with exposed thermal capacity.

Three alternative possible criteria are suggested for demonstrating compliance.

(1) For spaces with glazing facing one orientation only, limit the area of the glazed opening. The maximum area of glazed opening depends on the orientation and is expressed as a percentage of the area (measured internally) of the wall or roof element in which it occurs. Table 3.3 gives details.

Table 3.3   Maximum allowable areas of glazed opening.

Orientation of glazed opening	Maximum allowable area of opening %
North	50
North-east, north-west, south	40
East, south-east, west, south-west	32
Horizontal	12

(2)   For conditions of summer sunshine, show that the solar heat load per unit floor area, averaged between the hours of 07.30 and 17.30, does not exceed $25 \, \text{W/m}^2$. For this purpose, the condition of summer sunshine is defined as:

'The solar irradiances for the month of July at the location of the building that were not exceeded on more than 2.5% of occasions during the period 1976 to 1995.'

Strictly speaking, the solar irradiances vary with location, especially latitude. However, for the purposes of demonstrating compliance, Appendix H AD L2 (see Chapter 13) simplifies matters by giving a calculation procedure based on a single table that may be used for any location. This does not introduce a significant error over the range of latitudes to which Part L2 applies.

(3)   Show by detailed calculation procedures (e.g. CIBSE Guide A, Chapter 5 [38]), that in the absence of mechanical cooling or mechanical ventilation the space will not overheat when subjected to an internal gain of $10 \, \text{W/m}^2$. However, no criterion is provided to indicate when a space has become overheated.

3.3.2.3   Heating systems

Clearly, despite all the requirements in Schedule 1, the installation of combustion appliances and combustion systems (e.g. boiler plant and direct-fired gas heaters) must not infringe the necessary supply of air for health and for combustion. Reference must be made to:

- Part F   for the provision of adequate ventilation for health, *and*
- Part J   for the provision of adequate air for combustion appliances.

3.3.2.4   The carbon intensity of heating plant

The requirement is that heating plant should be reasonably efficient. The method of demonstrating compliance is to show that the amount of carbon generated by the heating plant is within specified limits at both maximum output and part load. The method applies to:

- Heating plant serving hot water and steam heating systems
- Electric heating
- Heat pumps (irrespective of the form of heat distribution).

The carbon intensity of the heat generating equipment must be calculated at maximum heat output of the heating system, *and* at 30% of this maximum, and the results must not exceed the values given in Table 3.4.

Table 3.4   Maximum allowable carbon intensities for heating systems.

Fuel	Maximum carbon intensities, kgC/kWh	
	At maximum heat output of the heating system	**At 30% of maximum heat output of the heating system**
Natural gas	0.068	0.065
Other fuels	0.091	0.088

The carbon intensity of the heating system depends on three factors:

- The rated output of each individual element of heat raising plant, R
- The gross thermal efficiency of each element of heat raising plant, measured in kWh of heat per kWh of delivered fuel, η_t
- The carbon emission factor of the fuel supplying each element of heat raising plant, measured in kg of carbon emitted per kWh of delivered fuel consumed, C_f.

The efficiency of consuming fuel in any equipment that supplies heat and power to the building is included in the gross thermal efficiency, the value of which depends on both the fuel and the equipment in which it is burnt. The carbon emission factor gives the total amount of carbon emitted and includes allowances for energy used in extracting, processing and delivering the fuel to the user. For example, electricity may consumed with very high efficiency by equipment within the building, but the generation of electricity from the primary fuel such as oil has a relatively low efficiency, and so losses have already occurred before the electricity itself is consumed. The same is true for other fuels. Values for the gross thermal efficiency must be obtained for each particular combination of equipment and fuel. However, values for C_f, the carbon emission factor, depend only on the fuel. A change in the production, manufacture or method of delivery of a fuel could cause the value of C_f to change, but any such change is unlikely to occur suddenly, and for the purposes of AD L2, C_f may be taken to be a constant for each fuel. The generally agreed values for C_f are shown in Table 3.5.

The carbon intensity, ε_c, is found by summing over all elements of heat generating plant:

Table 3.5 Carbon emission factors, C_f.

Delivered fuel	Carbon emission factor kgC/kWh	Comments
Natural gas	0.053	
LPG	0.068	
Biogas	0	
Oil	0.074	This value applies to all grades of fuel oil
Coal	0.086	
Biomass	0	
Electricity	0.113	Average for grid-supplied electricity, 2000–2005. For on-site generation by photovoltaics or wind power, use carbon emissions method. 'Green tarifs' etc. not appropriate
Waste heat	0	Includes waste heat from industrial processes, and from power stations of more than 10 MW electrical output and power efficiency better than 35%

$$\varepsilon_c = \frac{1}{\Sigma R}\Sigma\left(\frac{RC_f}{\eta_t}\right)$$

This equation is applicable to boilers, heat pump systems and electrical heating, but not combined heat and power (CHP) systems. In order to use the equation, data for the gross thermal efficiency for equipment must be obtained from the manufacturers or suppliers. For most practical cases, the efficiency may be taken as the efficiency at full load. However, where appropriate, a part load efficiency based on certified data supplied by the manufacturer may be used instead.

3.3.2.5 The carbon intensity of combined heat and power systems

A combined heat and power (CHP) system can create benefits because the on-site generation of electricity reduces the carbon emissions that would have occurred from the power stations feeding the national grid. This is taken into account by calculating the carbon intensity for CHP, ε_{chp}, from the equation:

$$\varepsilon_{chp} = \frac{C_f}{\eta_t} - \frac{C_{displaced}}{HPR}$$

where:

- η_t is the gross thermal efficiency of the CHP engine, measured in kWh of useful heat per kWh of fuel burned. Note that useful heat must be taken as the net amount supplied to the building after subtraction of any excess heat that the CHP system 'dumps' to the environment. For any given installation, it may be difficult to estimate the proportions of 'useful' and 'dumped' heat. It may therefore be necessary to demonstrate that η_t has been correctly evaluated, and AD L2 suggests the CHPQA certification scheme [39] as a means for doing this.

- C_f is the carbon emission factor of the fuel burned by the CHP engine, in kg of carbon emitted per kWh of delivered fuel consumed.
- $C_{displaced}$ is the carbon emission factor of the grid-supplied electricity which has been displaced by the CHP. Its value is that for the generating capacity which would otherwise be built if the CHP had not been provided, and is taken as 0.123 kgC/kWh (the value for new-generation gas-fired stations).
- HPR is the heat to power ratio of the CHP engine, in kWh of useful heat produced per kWh of electrical output. It is equivalent to the ratio of the thermal efficiency to the power efficiency of the CHP unit.

As for other heating systems, the carbon intensity for the CHP must be calculated for 100% and 30% of heating system output. If the CHP system operates alongside other heat generation equipment, then the carbon intensity for the complete system is found from:

$$\varepsilon_c = \frac{1}{\Sigma R} \Sigma \left(\frac{RC_f}{\eta_t} + R_{chp}\varepsilon_{chp} \right)$$

where R_{chp} is the rated output of the heat raising elements supplied by the CHP system.

3.3.2.6   *The carbon intensity of community heating and other heating methods*

The calculation of the carbon intensity of heat supplied to a building by a community heating system should take account of:

- The performance of the whole system, including the distribution circuits, all heat generating plant, and any CHP or waste heat recovery, *and*
- The carbon emission factors of all the different fuels.

The CHPQA Standard [39] provides a certification scheme for demonstrating that the thermal and power efficiencies have been estimated in a satisfactory manner.

In some buildings, especially factories, warehouses and workshops, local warm air or radiant heating systems may be acceptable and more efficient than centrally provided systems. Guidance on local systems is given in BRECSU publication GPG 303 [40].

3.3.2.7   Trade-off between construction elements and heating system efficiency

Throughout AD L2, rate of carbon emissions is taken to be an indicator of whether or not the requirement to conserve fuel and power has been met. Consequently, the elemental method allows the designer to trade off, in either direction, between the U-values of the building envelope and the carbon intensity of the heating system, provided the rate of carbon emissions is unchanged (or, although AD L2 does not specifically say, less). Compliance may be demonstrated by adjusting the area-weighted average U-value of the building fabric according to the equation:

$$U_{req} = U_{ref}\frac{\varepsilon_{ref}}{\varepsilon_{act}}$$

where U_{req} is the required maximum area-weighted average U-value of the building fabric

U_{ref} is the area-weighted average U-value of the building fabric when constructed to the elemental standards of Table 3.1

ε_{ref} is the carbon intensity of the reference heating system at an output of 30% of the installed design capacity, taken from Table 3.4 for the fuel type used in the actual heating system

ε_{act} is the carbon intensity of the actual heating system at an output of 30% of the installed design capacity.

Examples of the application of trade-off are given in Chapter 10.

3.3.2.8   Space heating controls

The building should be provided with controls that maintain the required temperature of each functional area *only during the period when it is occupied*. This means that the building and its heating system must have sufficient and appropriate zone, timing and temperature control devices. Additional controls may be provided to allow:

- Heating during extended unusual occupation hours
- Heating to prevent condensation or frost damage when the heating would otherwise be switched off.

AD L2 gives two ways of meeting the requirement:

- Buildings with a heating system maximum output of not more than 100 kW follow the guidance in BRECSU GPG 132 [41]
- Larger or more complex buildings follow the guidance in CIBSE Guide H [42].

Building control bodies may accept certification by a competent person that the requirements have been met.

3.3.2.9   Hot water systems and their controls

The requirement is to provide hot water safely while making efficient use of energy and thereby minimising carbon emissions. Ways of achieving this include:

- Avoidance of:
  (1) the over-sizing of hot water storage systems
  (2) low-load operation of heat raising plant
  (3) the use of grid-supplied electric water heating, except where demand is low.
- Provision of solar water heating
- Minimisation of
  (1) the length of circulation loops
  (2) the length and diameter of dead legs.

For conventional hot water storage systems, ways of satisfying the requirement would be to:

(1) Provide controls that shut off heating when the required water temperature is reached
(2) Shut off the supply of heat during periods when hot water is not required.

Guidance on ways of meeting the requirements include:

(1) In small buildings, Good Practice Guide GPG 132 [41]
(2) In larger, more complex buildings, or for non-conventional systems such as solar water heating, CIBSE Guide, Part H [42].

Building control bodies may accept certification by a competent person that the provisions meet the requirements.

3.3.2.10   Insulation of pipes and ducts

The requirements apply only to pipework, ductwork and vessels for the provision of:

- Space heating
- Space cooling, including chilled water and refrigerant pipework
- Hot water supply for normal occupancy.

The Building Regulations do not apply to pipework, ductwork and vessels for process use. The standards given in BS 5422 [29] are suitable for determining the amount of insulation to be applied to pipework, ductwork and storage vessels. In the case of storage vessels, the recommendations in BS 5422 [29] for flat surfaces should be used.

When, as a result of fluid flowing or being stored, the heat lost from pipes, ducts or vessels is always making a contribution to the conditioning of the surrounding space, insulation may not be necessary. Nevertheless, insulation may still be advisable to maintain stability and control of the fluid temperatures.

3.3.2.11 *Lighting efficiency standards*

AD L2 states that lighting systems should be reasonably efficient and, where appropriate, make effective use of daylight. Beyond this statement there is no further mention of daylight, and hence no guidance as to what level of daylight provision might be considered effective and acceptable. AD L2 is, however, concerned with conserving the energy used by electric lighting systems. The efficiency and the control of lighting systems are considered separately, and the guidance varies according to building type.

3.3.2.11.1 General lighting efficacy in office, industrial and storage buildings
It should first be noted that the definition of luminous efficacy used here differs from that used for domestic lighting in section 2.3.8. Thus:

$$\text{Luminous efficacy} = \frac{\text{Total lumens emitted by lamp and luminaire combined}}{\text{Total circuit-watts consumed by lamp}}$$

With this definition in mind, the electric lighting system should be provided with reasonably efficient lamp/luminaire combinations. The requirement can be met by a luminous efficacy, averaged over the whole building, of not less than 40 luminaire-lumens per circuit-watt.

When interpreting this criterion, it should be noted that:

- The figure of 40 refers to the number of lumens emitted by the lamp and luminaire combination, and not just the lamp.
- The circuit-watts includes all the power consumed in the lighting circuits, including the lamps, their associated control gear and power factor correction equipment.
- The effect of the above is to allow flexibility in the choice of lamp, luminaire and control gear. For example, a low efficiency luminaire (chosen perhaps for reasons of appearance or low glare) can be compensated by a lamp of high efficacy and/or more efficient control gear.
- A maximum of 500 W of installed lighting may be excluded from this guidance and hence from the calculations. This is to allow flexibility for the use of feature lighting, etc.

The initial luminaire efficacy, η_{lum}, in luminaire-lumens per circuit-watt may be calculated by means of the equation:

$$\eta_{lum} = \frac{1}{P} \sum \left(\frac{\text{LOR} \times \phi_{lamp}}{C_L} \right)$$

where   LOR is the light output ratio of the luminaire. The LOR is defined as the ratio of the total light output of the luminaire under stated practical conditions to the total light output of the bare lamp or lamps used in the luminaire under reference conditions

ϕ_{lamp} is the sum of the average initial (100 hour) lumen output of all the lamps in the luminaire

P is the total circuit-watts for all luminaires

C_L is a control factor which takes account of controls which reduce the output of the luminaire when electric light is not required. The control factor, C_L, is an empirical parameter which allows for the energy savings which are expected to accrue on average from lighting control strategies. It is based in part on observations of occupant behaviour, and cannot be derived from first principles. Some values are given in Table 3.6.

Table 3.6   Luminaire control factors.

Control function	C_L
The luminaire is in a daylit space* and its light output is controlled by: *either* a photoelectic switching or dimming control, with or without manual override *or* local manual switching**	0.80
The luminaire is in a space that is likely to be unoccupied for a significant proportion of working hours and where a sensor switches off the luminaires in the absence of occupants but switching on is done manually	0.80
Both of the above circumstances combined	0.75
None of the above	1.00

Definitions
* *Daylit space:* Any space within 6 m of a window wall provided that the glazed area is at least 20% of the internal area of the window wall or a roof-lit space with a glazing area at least 10% of the floor area. The glazing must have a light transmittance at normal incidence (i.e. to light at an angle perpendicular to its surface) of at least 70%, or to a lower value if the glazed area is increased in proportion to its area (provided that the maximum areas to limit heat loss and to avoid solar overheating are not exceeded).
** *Local manual switch:* A switch whose distance from the luminaire it controls is not more than 8 m on plan, or three times the height of the luminaire above the floor if this is greater *and* is operated by deliberate action of the occupants (rocker switch, push button, pull cord, etc.) *or* is operated by remote control (infrared transmitter, sonic or ultrasonic devices, telephone handset controls, etc.).

3.3.2.11.2   General lighting efficacy in all other building types

In other building types, it may be appropriate to use luminaires for which photometric data is not available (i.e. the value of the LOR is not known), and/ or luminaires which take low power less efficient lamps. For these cases, the criterion is based on the light output of the lamps themselves, without the need to allow for the performance of the luminaire. That is to say, the definition of luminous efficacy corresponds to that used for domestic lighting. The requirement is that the installed lighting capacity has an initial (100 hour) lamp plus ballast efficacy of not less than 50 lamp-lumens per circuit-watt. This would be taken as achieved if at least 95% of the installed lighting capacity used lamps as described in Table 3.7.

Table 3.7   Light sources which meet the criterion for general lighting.

Light source	Types and ratings
High pressure sodium	All types and ratings
Metal halide	All types and ratings
Induction lighting	All types and ratings
Tubular fluorescent	38 mm diameter (T12) linear fluorescent lamps 2400 mm in length
Tubular fluorescent, with high efficiency control gear*	26 mm diameter (T8) lamps, and 16 mm diameter (T5) lamps rated above 11 W
Compact fluorescent	All ratings above 11 W
Other	Any type and rating with an efficacy greater than 50 lamp-lumens per circuit-watt

* *High efficiency control gear:* This means low loss or high frequency control gear that has a power consumption, including the starter component, not exceeding a specified value. The specified value depends on the nominal lamp rating and may be obtained from Table 3.8.

Table 3.8   Maximum power consumption of high efficiency control gear.

Nominal lamp rating, watts	Maximum power consumption, watts
Less than or equal to 15	6
Greater than 15, not more than 50	8
Greater than 50, not more than 70	9
Greater than 70, not more than 100	12
Greater than 100	15

3.3.2.11.3   Display lighting for all buildings

Within AD L2, display lighting is defined as:

- Lighting intended to highlight displays of exhibits or merchandise
- Lighting used in spaces for public entertainment (e.g. dance halls, auditoria, conference halls and cinemas).

The special requirements of display lighting may make it necessary to accept lower standards of energy performance than for general lighting. Nevertheless, it is still necessary for display lighting to be energy efficient, and AD L2 offers two ways of demonstrating compliance:

- *either* ensure that the installed capacity of the display lighting has an initial (100 hour) efficacy of not less than 15 lamp-lumens per circuit-watt
- *or* ensure that at least 95% of the installed display lighting capacity in circuit-watts is made up of lamps and fittings that have circuit efficacies no worse than the following:
  - ○ High pressure sodium   all types and fittings
  - ○ Metal halide   all types and fittings
  - ○ Tungsten halogen   all types and fittings
  - ○ Compact and tubular fluorescent   all types and fittings.

The circuit-watts should include the power consumed by transformers and ballasts.

3.3.2.11.4   Emergency escape lighting and specialist process lighting
Lighting for these purposes is not subject to the requirements of Part L. These types of lighting are defined as:

- *Emergency escape lighting:* that part of emergency lighting that provides illumination for the safety of people leaving an area or attempting to terminate a dangerous process before leaving an area.
- *Specialist process lighting:* Lighting that is intended to illuminate specialist tasks within a space rather than the space itself.

Specialist process lighting could include:

- Theatre spotlights
- Projection equipment
- Lighting in TV and photographic studios
- Medical lighting in operating theatres and doctors' and dentists' surgeries
- Illuminated signs
- Coloured or stroboscopic lighting
- Integral lighting for art objects such as sculptures
- Decorative fountains and chandeliers.

3.3.2.12   Lighting controls

The aim of lighting controls should be to encourage the maximum use of daylight and to avoid unnecessary use of lighting when spaces are not occupied. This should not, however, create a situation in which the operation of an automatically switched lighting system endangers the passage of building occupants. Guidance on lighting controls is given in BRE IP 2/99 [43].

3.3.2.12.1   Controls in offices and storage buildings

The requirement can be met by providing local switches in easily accessible positions within each working area or at boundaries between working areas and circulation routes. In the context of AD L2, a switch is taken to include a dimmer switch, and switching includes dimming. However, as a general rule, dimming should be achieved by reducing rather than by diverting the energy supply. Local switches could include:

- A switch that can be operated by the deliberate action of an occupant either manually or by remote control, *and/or*
- Automatic switching systems that switch off the lighting when a sensor senses the absence of occupants.

Local switches include:

- Rocker switches, push buttons and pull cords operated manually, *and/or*
- Infrared, sonic and ultrasonic transmitters, and telephone handset controls, operated remotely.

Local switching can be supplemented by other controls such as timers and photo-electric devices.

The positioning of local switches is important. The distance on plan from any local switch to the luminaire it controls should not normally be more than 8 m, or three times the height of the luminaire above the floor if this is greater.

3.3.2.12.2   Controls in buildings other than offices and storage buildings

Again, the designer is urged to maximise the beneficial use of daylight without being given any specific criterion for daylight provision. The guidance is limited to a requirement to use, as appropriate, strategies to ensure that artificial lighting is switched off whenever daylight levels within a building are sufficiently high. The lighting controls should be one or more of:

- Local switching (see above)
- Time switching (e.g. in major occupational areas which have clear timetables of occupation)
- Photo-electric switching.

3.3.2.12.3   Controls for display lighting

Connect display lighting in dedicated circuits that can be switched off when not required. For example, in a retail store, timers could be used that switch off display lighting outside store hours, except for displays intended to be viewed from outside through display windows.

3.3.2.13 Air conditioning and mechanical ventilation (ACMV)

Within the context of AD L2, the acronym ACMV includes mechanical ventilation systems with no cooling facility as well as mechanical ventilation systems with full air conditioning. Definitions are provided for the terms:

- mechanical ventilation
- air conditioning
- treated areas
- process requirements.

Mechanical ventilation: This is a system that uses fans to supply outdoor air and/or extract indoor air to meet ventilation requirements. The system may be extensive and may include components such as air handling units, air filtration and heat reclamation. The system does *not* provide active cooling from refrigeration equipment. The definition does *not* apply to naturally ventilated buildings which use individual extract fans (mounted in either a wall or a window) to improve the ventilation of a small number of rooms.

Air conditioning: This is any system where refrigeration is included to provide cooling for the comfort of the building's occupants. The cooling function can be provided from stand-alone refrigeration equipment in the cooled space, or from centralised or partly centralised equipment, or from systems that combine cooling with mechanical ventilation.

Treated area is the floor area, measured between the internal faces of the surrounding walls, of the spaces that are served by a mechanical ventilation or air conditioning system. Treated area does *not* apply to spaces (plant rooms, service ducts, lift wells, etc.) which are *not* served by a mechanical ventilation or air conditioning system; such spaces should be excluded from the treated area total.

Process requirements: In an office building, process requirements include any significant area within which an activity takes place that is not typical of an ordinary commercial office. The performance of the mechanical ventilation and air conditioning systems for such areas are determined by the process requirements, and these areas, together with the plant capacity (or proportion of plant capacity) associated with them, should be excluded from calculations. In office buildings, activities and areas that may be considered to fall within the definition of process requirements include:

- Staff restaurants and kitchens
- Large dedicated conference rooms
- Sports facilities
- Dedicated computer or communications rooms.

The general requirement for ACMV is that buildings should be designed and constructed such that:

- The form and fabric of the building do not create a need for excessive installed ACMV capacity. Glazing type, glazing ratios, and solar shading are important in limiting cooling requirements
- Fans, pumps and refrigeration equipment are reasonably efficient and are not over-sized, so that the capacity for demand and standby is no more than necessary
- Facilities for the management, control and monitoring of the operation of equipment and systems are provided.

The method of demonstrating compliance depends on the building type. For buildings which are used as offices, or are of a similar type and usage to an office development, the carbon performance rating (CPR) method [44] can be used. For all other building types the only criterion is the specific fan power (SFP) of the mechanical systems.

3.3.2.13.1   The CPR method for office buildings with ACMV
The CPR method can be used if there are no innovative building or building services provisions. Otherwise, the carbon emissions calculation method, or some other acceptable alternative, must be used. The formulae and method of calculation of the CPR are given in Appendix G of AD L2 (see Chapter 12). Compliance is demonstrated if the CPR does not exceed the relevant maximum value given in Table 3.9. If part of a building is served by a new air conditioning system and part served by a new mechanical ventilation system, the two parts should be considered separately and the relevant CPR must be met in each part.

Table 3.9   Maximum allowable carbon performance ratings.

	Maximum CPR, kgC/m²/year	
System type	New building	Existing building
Air conditioning	10.30	11.35
Mechanical ventilation	6.50	7.35

Where substantial alteration is made to an existing ACMV system, compliance would be achieved if:

- *Either* the CPR is reduced by at least 10% by the work
- *Or* the new CPR does not exceed the relevant value in Table 3.9.

Where there is only replacement of existing equipment, the criterion is the product of the installed capacity per unit area, PD (or PR), and the control management factor, FD (or FR). (See Appendix G of AD L2 for the full definitions of PD, etc.). The product PD × FD (or PR × FR) should:

- *Either* be reduced by at least 10%
- *Or* meet a level of performance equivalent to the component benchmarks given in CIBSE TM 22 [45].

3.3.2.13.2   Methods for other buildings with ACMV

Other mechanically ventilated buildings may be assessed for compliance by means of the specific fan power (SFP). This is the sum of all the circuit-watts used to drive all the supply and extract ventilation fans, including switchgear, inverters, etc., divided by the design ventilation rate, in litres per second, of the building. The SFP may be used regardless of whether or not the air supply is heated or cooled. For typical spaces ventilated for human occupancy, compliance is demonstrated if:

- for ACMV systems in new buildings, the SFP is no greater than 2.0 W/litre/ second
- for new ACMV in refurbished buildings, or where an existing ACMV system is being substantially altered, the SFP is no greater than 3.0 W/litre/second.

For spaces where higher ventilation rates are required because of, say, specialist processes or higher than normal external pollution levels, higher values of SFP may be appropriate.

It is also important that mechanical ventilation systems are reasonably efficient at part load. This could be demonstrated by providing efficient variable flow control systems such as variable speed drives or variable pitch axial flow fans. Detailed guidance is given in BRESCU GIR 41 [46].

3.3.3   The whole-building method

This method is separate and independent of the elemental method and therefore allows much more design flexibility. To show compliance it must be shown that:

- *Either* the total carbon emissions
- *Or* the primary energy consumption

for the complete building are reasonable for the purposes of conserving fuel and power. Three building types are considered in AD L2, though for all three types the details and calculation procedures are given not in AD L2 but in other publications.

3.3.3.1   Office buildings

Office buildings can be treated by means of the whole-office carbon performance rating method. In principle this is the same as the CPR method for assessing the ACMV systems of office buildings (described in section 3.3.2.13.1 above), and which is part of the elemental method. However, when used as part of the elemental method, the CPR calculation deals only with the building's ACMV systems, whereas in the whole-building method, the CPR calculation is expanded to include lighting and space heating. Full details are to be found in BRE Digest No. 457 [44]. There are several assumptions implicit in using the method:

- The proposed design of the building and its energy consuming systems are capable of creating internal environmental conditions which are normal and acceptable for the occupants and the functional requirements of the building. Otherwise, it would be possible to meet the relevant criteria by deliberately undersizing the installed equipment.
- The design of the ACMV, space heating and lighting systems are conventional and do not require the use of novel equipment. This means that the heating system is likely to be supplied from a conventional boiler, and that cooling will be provided by conventional refrigeration plant.
- The CPR method applies only to the energy consuming systems themselves. Therefore, while it provides the flexibility for adjustment between the energies consumed by the ACMV, the heating system and the lighting system, the method is not intended to provide an excuse for poor fabric design.
- The CPR calculation is intended to be a relatively simple method for assessing a building and its systems for compliance. If certain features of the design are likely to make the calculation unduly complex, one of the alternative methods should be used.

The question of whether or not particular equipment may be considered conventional and within the scope of the method must be carefully considered. In some cases, it depends on the detail:

- *Heat recovery* can be included if the effect is to reduce the size of the heat generator. It can also be included, though not so easily, if the effect is to reduce the hours of use of the heat generator.
- *Thermal storage* can be included for ice thermal storage which is part of a cooling/refrigeration system. For thermal stores whose purpose is to even out the peaks in demand on heat generators, the CPR method is not recommended.
- *Space heating using heat pumps* is possible but not advised. The designer could not use the factors given as part of the method, and would have to make his own calculation of the carbon emissions.
- *Combined heat and power* is also possible but not advised. The calculation of the carbon emissions could be excessively complex.
- *Renewable energy* is possible in some cases. The recommended technique is to calculate the carbon emissions assuming all the energy is supplied from conventional sources, and then to calculate the saving due to the renewable energy source. The amount saved is then subtracted from the initial result.

The calculation procedure is described in Chapter 12. When the CPR value of an office building has been found, it can be considered to comply if it meets *all* the following:

- A whole-office CPR which is no worse than the relevant maximum in Table 3.10
- The requirements for avoiding thermal bridging at junctions and around openings (as described previously)
- The requirements for meeting air leakage standards
- The upper limits for U-values, i.e for parts of a roof $U \leq 0.35\,\mathrm{W/m^2K}$, and for parts of an exposed wall or floor $U \leq 0.70\,\mathrm{W/m^2K}$.

Table 3.10   Maximum whole-office CPR.

Office building type	Maximum allowable CPR, kgC/m²/year	
	New office	Refurbished office
Naturally ventilated	7.1	7.8
Mechanically ventilated	10.0	11.0
Air-conditioned	18.5	20.4

3.3.3.2   *Schools*

For schools, compliance may be demonstrated by showing that the building conforms with DfEE Building Bulletin 87 [47].

3.3.3.3   *Hospitals*

For hospitals, compliance may be demonstrated by showing that the building conforms with the NHS Estates Guide [48].

3.3.4   Carbon emissions calculation method

The carbon emissions method allows considerable flexibility in the design of a building. The design may take advantage of any energy conservation measure and may take into account heat gains due to solar radiation and internal heat sources. To show compliance using the carbon emissions method, four conditions must be met:

- The calculated annual carbon emissions of the proposed building should be no greater than from a notional building of the same size and shape which has been designed to comply with the elemental method
- For the notional building, the U-value of the floor must be taken as *either* $0.25\,\mathrm{W/m^2K}$ *or* the U-value of the floor in the proposed building, whichever is the lower

- For the proposed building, the poorest acceptable U-values for parts of the roof and walls are 0.35 W/m^2K for the roof and 0.70 W/m^2K for the walls
- The fabric of the proposed building must be constructed to provide resistance to air leakage to at least the same standard as required by the elemental method.

A critical feature of the carbon emissions method is the procedure which is used to carry out the calculation. In general, simple methods which do not require the aid of a computing device are very unlikely to be sufficient, and so in nearly all cases the calculations must be performed by computer using either in-house or commercially purchased software. The reliability of the method being used is therefore a major consideration, and the Approved Document requires that the method is acceptable without specifying those methods which would be acceptable. Tests of acceptability are:

- That the method has been approved by a relevant authority responsible for issuing professional guidance
- The organisation responsible for carrying out the calculations is using a method which satisfies their own in-house quality assurance procedures
- Either of the above can be demonstrated by submitting with the calculations a completed copy of Appendix B of CIBSE AM11 [49]; this is a checklist for choosing BEEM software, and should show that the software which has been used is appropriate for the purpose.

3.3.5   Other matters – conservatories, atria, sun-spaces, etc.

3.3.5.1   Definitions

Sun-space (including conservatory and atrium): A sun-space is a building or part of a building having not less than three-quarters of the area of its roof and not less than half the area of its external walls (if any) made of translucent material.

Separation: Separation between a building and a sun-space means:

- The separating walls and floors are insulated to at least the same degree as the exposed walls and floors of the building
- The separating windows and doors have U-values and draught proofing to at least the same standard as the windows and doors elsewhere in the building.

3.3.5.2   A sun-space attached to and built as part of a new building

Where there is *no separation* between the sun-space and the building, the sun-space should be treated as an integral part of the building.

Where there is *separation* between the sun-space and the building, energy savings can be achieved if the sun-space is neither heated nor mechanically cooled. If fixed heating or mechanical cooling is installed, they should have their own separate temperature and on/off controls.

3.3.5.3 *A sun-space attached to an existing building*

When attaching a sun-space to an existing building, reasonable provision should be made to limit heat loss from, or summer solar heat gain to, the building. Ways of meeting this requirement are:

- If the opening is not to be enlarged, retain the existing separation, *or*
- If an opening has to be enlarged or newly created as a material alteration, provide separation as or equivalent to windows and doors having the average U-value given in Table 3.1 for the elemental method.

3.4 AD L2 – Section 2 Construction

The persons or organisations who construct and assemble the building have to satisfy the building control body of a number of matters, two of which are dealt with in this section:

- Certain aspects of the building fabric
- Inspection and commissioning of the building services systems.

3.4.1 Building fabric

3.4.1.1 *Continuity of insulation*

Continuity of insulation is necessary in order to avoid excessive thermal bridging. The design requirements for achieving this are as described for the elemental method in section 3.3.2.2.1. In addition, the person carrying out the work has a responsibility to ensure that compliance with Part L is achieved. For a new building, that person will normally be:

- The developer who has carried out the work subject to Part L
- The contractor who has carried out the work subject to Part L
- The sub-contractor who has carried out the work subject to Part L.

If he is suitably qualified, the person responsible for achieving compliance should provide a certificate or declaration stating that the requirements of Part L2(a) have been met. Otherwise, a certificate or declaration to that effect should be obtained from a suitably qualified person. The certificate/declaration must be based on:

- *Either* a statement confirming that the design details, building techniques and manner in which the work has been carried out can be expected to achieve reasonable conformity with specifications that have been approved for compliance with Part L2
- *Or* post-completion testing using infrared thermography which shows that the fabric insulation is reasonably continuous over the whole visible envelope. Information on thermography for building surveys is given in BRE Report 176 [50].

3.4.1.2   Airtightness

Details of the requirements for minimising air infiltration through the building fabric are given under the elemental method in section 3.3.2.2.2, together with the standards that must be achieved. In addition, the person carrying out the building work must obtain certificates or declarations that confirm that the required standard of airtightness has been achieved:

- *Either* for buildings of any size, by means of air leakage tests carried out in accordance with CIBSE TM 23 [21]
- *Or* for buildings of less than $1000\,\text{m}^2$ gross floor area, by using appropriate design details and building techniques, and by carrying out the work in ways that should achieve reasonable conformity with the specifications that have been approved for compliance with Part L2.

3.4.1.3   Certificates and testing

Certificates/declarations such as those described in sections 3.4.1.1 and 3.4.1.2 above may be accepted by building control bodies as evidence of compliance. However, it is necessary to establish, to the satisfaction of the building control body and in advance of the work, that the person who will give the certificates/ declarations is suitably qualified.

3.4.2   Inspection and commissioning of building services systems

When describing or discussing the requirements for building services systems in Part L2, the terms 'providing' and 'making provision' include, where relevant, inspection and commissioning, the definitions of which are as follows.

3.4.2.1   Inspection of building services systems

This is defined as the establishment, at completion of installation, that the specified and approved provisions for efficient operation have been put in place.

3.4.2.2   Commissioning of building services systems

This means the advancement of these systems from the state of static completion to working order to the specifications relevant to achieving compliance with

Part L2, without prejudice to the need to comply with health and safety requirements. For each system, this includes:

- Setting-to-work
- Regulation, i.e. testing and adjusting repetitively to achieve the specified performance
- Calibration, setting up and testing of the associated automatic control systems
- Recording of the system settings and the performance test results that have been accepted as satisfactory.

Responsibility for achieving compliance with the requirements of Part L lies with the person carrying out the work. For building services systems, this person may be:

- The developer who has carried out the work subject to Part L
- The main contractor who has carried out the work subject to Part L
- A sub-contractor who has carried out the work subject to Part L
- A specialist firm directly engaged by a client.

A report must be provided, either by the person responsible for achieving compliance or by a suitably qualified person, that indicates that the inspection and commissioning activities necessary to establish that the work complies with Part L have been completed to a reasonable standard. The report should include:

- A commissioning plan that shows that every system has been inspected and commissioned in an appropriate sequence
- The results of the tests that confirm that the performance is in reasonable accordance with the approved designs, including written commentaries where excursions are proposed to be accepted.

Compliance may be demonstrated by following the guidance in the CIBSE Commissioning Codes [51] and TM1 of the Commissioning Specialists Association [52].

3.4.2.3 Reports and testing

A report such as that described above may be accepted by building control bodies as evidence of compliance. However, it is necessary to establish, to the satisfaction of the building control body and in advance of the work, that the person who will provide the report is suitably qualified.

3.5   AD L2 – Section 3 Providing Information

There is an obligation to provide the owner and/or the occupier of the building with certain information. There are two aspects to this: the provision of a building log-book and the installation of energy meters.

3.5.1   Building log-book

The building owner and/or occupier must be provided with a log-book. The log-book should contain details of:

- The installed building services plant
- The installed building services controls
- The method of operation of the plant and controls
- Maintenance requirements
- Any other matters which collectively enable energy consumption to be monitored and controlled.

The log-book information should be provided in summary form and be suitable for use on a day-to-day basis. The log-book may refer to information contained in other documents, for example, operation and maintenance manuals, health and safety files, etc.

Log-book contents
The log-book could include the following:

(1)   A description of the whole building, including its intended usage and the philosophy of the design
(2)   A description of the intended purpose of the individual building services systems
(3)   A schedule of the floor areas of each building zone, broken down according to the type of environmental service provided to that zone, i.e. air conditioning, natural ventilation, etc.
(4)   The location of all relevant plant and equipment, including simple schematic diagrams
(5)   The installed capacity of the services plant, expressed as the input power and the output rating
(6)   Simple descriptions of the proper strategies for operating and controlling the energy consuming services of the building
(7)   A copy of the report that confirms that the building services equipment has been commissioned and found to be satisfactory
(8)   Inclusion, in the operating and maintenance instructions, of the provisions that enable the specified performance to be sustained during occupation

(9)   A schedule of the building's energy supply meters and sub-meters; for every meter and sub-meter, this schedule should include:
   • location
   • identification and description
   • the type of fuel being monitored
   • instructions on its use – these instructions should indicate how the energy performance of the building (or, if relevant, each separate tenancy in the building) can be calculated from the metered energy readings for comparison with published benchmarks (see also Appendix G of AD L2, and see the next section below for metering strategies)

(10)   For systems serving an office floor area of more than $200\,m^2$, a design assessment of the building services systems' carbon emissions and the comparable performance benchmark (see Appendix G of AD L2)

(11)   The measured air permeability of the building.

3.5.2   Installation of energy meters

The building engineering services should be provided with sufficient energy meters and sub-meters to enable owners or occupiers to measure their actual energy consumption. Sufficient instructions, including an overall metering strategy, must be provided so that owners or occupiers are able to attribute energy consumed to the end use of that energy, and to be able to compare operating performance to published benchmarks (see item 9 of section 3.5.1.2 above). In order to develop a metering strategy, it is first necessary to know how the energy from each fuel will be used in the building, and to have a good estimate of the amount of each fuel that will be consumed. CIBSE TM22 [45] provides a standardised procedure for doing this, including prepared spreadsheets and a worked example on CD-ROM. The procedure is in three stages:

• Stage 1 is a quick assessment in terms of energy use per unit floor area
• Stage 2 is an improved assessment accounting for special energy uses, occupancy and weather
• Stage 3 is a detailed assessment of the building and all its energy systems.

Once the information on energy usage has been collected, a metering strategy can be developed. Detailed guidance on metering strategies, including worksheets and worked examples, is given in GIL 65 [53].

3.5.2.1   *Reasonable provision for energy metering*

Provision for energy metering is considered to be reasonable when it is possible to measure at least 90% of the estimated annual energy consumption of each fuel to be accounted for. Possible techniques by which the allocation of energy consumption to each end use can be achieved are:

- Direct metering
- Measuring the run-hours of equipment that operates at a constant known load
- Estimating the energy consumption indirectly; for example, this could be done for hot water supply by combining measurements of the amount of water supplied and the delivery temperatures of the water, together with the known efficiency of the water heater
- Estimating the energy consumption by difference; for example, by measuring the total consumption of gas and deducting the measured gas consumption for heating and hot water, the amount of gas used for some other function (e.g. catering) can be found
- Estimating non-constant small power loads by means of the procedure given in CIBSE Energy Efficiency Guide, Chapter 11 [36].

3.5.2.2   Reasonable provision of energy meters and sub-meters

Reasonable provision of meters would be to install incoming meters in every building greater than 500 m² gross floor area (including separate buildings on multi-building sites). This would include:

- Individual meters for direct measurement of the total electricity, gas, oil and LPG consumed by a building
- A heat meter capable of direct measurement of the total heating and/or cooling energy supplied to the building by a district heating or cooling scheme.

In the case of sub-metering, it would be reasonable to provide additional meters to directly measure or reliably estimate (see section 3.5.2.1 above):

- The electricity, natural gas, oil and LPG supplied to each separately tenanted area that is greater than 500 m²
- The energy consumed by plant items with input powers greater than or equal to those listed in Table 3.11
- Any heating or cooling supplied to separately tenanted spaces: for tenancies of floor area greater than 2500 m², direct metering of the heating and cooling may be appropriate, but for smaller tenanted areas, the heating and cooling end uses may be proportioned on an area basis
- Any process load (see 'Process requirements' in section 3.3.2.13) that is discounted from the building's energy consumption when comparing measured consumption against published benchmarks.

Table 3.11 Size of plant for which separate metering would be reasonable.

Plant item	Rated input power, kW
Boiler installations comprising one or mroe boilers or CHP plant feeding a common distribution circuit	50
Chiller installations comprising one or more chiller units feeding a common distribution circuit	20
Electric humidifiers	10
Motor control centres providing power to fans and pumps	10
Final electrical distribution boards	50

3.6   AD L2 – Section 4 Work on Existing Buildings

3.6.1   Replacement of a controlled service or fitting

Attention must be paid to the definitions of 'controlled service or fitting' and 'building work'. These are the same as for dwellings and are given in section 2.4.1. The requirement to make reasonable provision applies when:

- Replacing old with new identical equipment, *or*
- Replacing old with different equipment.

It also applies:

- When the work is solely in connection with controlled services, *or*
- When the work includes work on controlled services.

Ways of meeting the requirements include the following.

3.6.1.1   Windows, doors and rooflights

When these elements are replaced:

- Provide units that meet the requirements for new buildings, *or*
- Provide units with a centre-pane U-value no worse than $1.2\,W/m^2K$.

The replacement work should comply with the requirements of both Part L2 and (unless non-glazed fittings are involved) Part N. In addition, after the work the building should not have a worse level of compliance with other relevant parts of Schedule 1, such as Parts B, F and J.

  However, note that the requirement does not apply to repair work on parts of these elements, such as replacing broken glass, sealed double glazing units or rotten framing members.

3.6.1.2   Heating systems

If the heating system is substantially replaced, then provide a new heating system and new controls to the standards required for new installations (i.e. a new building). In lesser work, it is acceptable to provide insulation, zoning, timing, temperature and interlock controls.

Again, when meeting other relevant requirements of Schedule 1, particular account should be taken of Parts F and J.

3.6.1.3   Hot water systems

When substantial replacement to hot water systems, pipes and vessels takes place, provide controls and insulation as if for a new building. In lesser work, it is sufficient to provide insulation, and timing and thermostatic controls.

3.6.1.4   Lighting systems

If a complete lighting system serving more than $100 \, \text{m}^2$ of floor area is to be replaced, then the new system should comply with the requirements for a new building. In the case of partial replacement, then:

- If only the complete luminaries are replaced, provide new luminaries as specified in either section 3.3.2.11.1 or section 3.3.2.11.2, whichever is appropriate (note however that this requirement does not apply when only components such as lamps or louvres are replaced)
- If only the control system is to be replaced, provide new controls as specified in either section 3.3.2.12.1 or section 3.3.2.12.2, whichever is appropriate (note however that this requirement does not apply when only components such as switches or relays are replaced).

3.6.1.5   Air conditioning or mechanical ventilation systems

For office buildings, when replacing systems which serve more than $200 \, \text{m}^2$ of floor area, the carbon performance rating should be improved to the standards given in section 3.3.2.12.1.

For all other buildings, provide mechanical ventilation systems that meet the SFP requirements of section 3.3.2.13.2.

3.6.1.6   Commissioning, etc.

When carrying out any of the work described in sections 3.6.1.2 to 3.6.1.5 inclusive:

- The work should be inspected and commissioned as described in section 3.4.2

- A building log-book should be prepared, or the existing log-book updated, with details of the replacement controlled service or fitting, as described in section 3.5.1
- In order that the replacement controlled service or fitting can be effectively monitored (as described in section 3.5.2), the relevant part of the metering strategy should be prepared or revised as necessary, and additional metering provided where needed.

3.6.2   Material alterations

Attention must be paid to the definitions of 'material alteration' and 'relevant requirement'. These are the same as for dwellings and are given in section 2.4.2. When undertaking material alterations, account should be taken of:

- All the relevant requirements of Schedule 1, including Parts L2, F and J
- Insulation of roofs, floors and walls
- Sealing measures
- Controlled services and fittings.

When undertaking material alterations, reasonable provision for satisfying the requirements of Part L2 may depend on the circumstances of the particular case, and would need to take into account any historic value in the structure being altered. Specific guidance is as follows.

3.6.2.1   Roof insulation

If the material alteration includes substantial replacement of any of the major elements of a roof structure, insulate to the U-value standard required for a new building.

3.6.2.2   Floor insulation

If the structure of a ground floor is to be substantially replaced or re-boarded, and if the room is heated, insulate to the U-value standard of a new building.

3.6.2.3   Wall insulation

Provide a reasonable thickness of insulation when substantially replacing:

- Complete exposed walls
- The external rendering or cladding of an exposed wall
- The internal surface finishes of an exposed wall
- The internal surfaces of separating walls to unheated spaces.

3.6.2.4   *Sealing measures*

When carrying out any of the above work on roofs, floors or walls, include reasonable sealing measures to improve airtightness.

3.6.2.5   *Controlled services or fittings*

Follow the guidance in section 3.6.1 above.

3.6.3   Material changes of use

The definition of a material change of use given in section 2.4.3 for dwellings also applies to other buildings. When undertaking material alterations, account should be taken of:

- All the relevant requirements of Schedule 1, including Parts L2, F and J
- Insulation of accessible lofts, roofs, floors and walls
- Sealing measures
- Controlled services and fittings.

When undertaking material alterations, reasonable provision for satisfying the requirements of Part L2 may depend on the circumstances of the particular case, and would need to take into account any historic value in the structure which is subject to the material alteration. Specific guidance is as follows.

3.6.3.1   *Accessible lofts*

Where the existing insulation in accessible lofts is worse than $0.35 \, \text{W}/\text{m}^2\text{K}$, replace or add extra insulation to upgrade the U-value to a maximum of $0.25 \, \text{W}/\text{m}^2\text{K}$.

3.6.3.2   *Roof insulation*

If the material alteration includes substantial replacement of any of the major elements of a roof structure, insulate to the U-value standard of a new building.

3.6.3.3   *Floor insulation*

If the structure of a ground floor is to be substantially replaced, and if the room is heated, insulate to the U-value standard of a new building.

3.6.3.4   *Wall insulation*

Provide a reasonable thickness of insulation when substantially replacing:

- Complete exposed walls
- The external rendering or cladding of an exposed wall
- The internal surface finishes of an exposed wall
- The internal surfaces of separating walls to unheated spaces.

3.6.3.5   *Sealing measures*

When carrying out any of the above work on roofs, floors or walls, include reasonable sealing measures to improve airtightness.

3.6.3.6   *Controlled services or fittings*

Follow the guidance in section 3.6.1 above.

3.6.4   Historic buildings

Historic buildings include:

- Listed buildings
- Buildings situated in conservation areas
- Buildings of architectural and historical interest and which are referred to as a material consideration in a local authority development plan
- Buildings of architectural and historical interest within national parks, areas of outstanding natural beauty, and world heritage sites.

Any work on an historic building must balance the need to improve energy efficiency against the following factors:

- The need to avoid prejudicing the character of the historic building
- The danger of increasing the risk of long-term deterioration of the building fabric
- The danger of increasing the risk of long-term deterioration of the building's fittings
- The extent to which energy conservation measures are a practical possibility.

Information on the special characteristics of historic buildings and their conservation should be obtained, for example from Planning Policy document PPG15 [33] and BS 7913 [34]. Advice on achieving the correct balance should also be sought from the conservation officer of the local authority. Advice from other sources would may also be appropriate, particularly regarding:

- Restoration of the historic character of a building that has been the subject of inappropriate alteration, such as the replacement of windows, doors or rooflights

- Rebuilding of a former historic building, which may have been damaged or destroyed due to some mishap (such as a fire), or infilling a gap in a terrace
- Providing a means for the fabric of an historic building to 'breathe' so that moisture movement may be controlled and the potential for long-term decay problems reduced (see SPAB Information Sheet No. 4 [35]).

4 Tables of U-values

Appendix A of the Approved Documents provides look-up tables for obtaining U-values for a range of windows, doors and rooflights. For all other elements, it provides look-up tables which allow the user to find, by relatively simple calculations, the minimum thickness of insulation required to meet a specified U-value.

4.1 Windows, doors and rooflights

When available, manufacturers' certified U-values (by approved methods of measurement or calculation) should be used. If these are not available, values for single, double and triple glazing may be taken from Tables 4.1, 4.2 and 4.3, modified where necessary for metal frames according to Table 4.4. Low emissivity (low-E) coatings are of two main types, 'hard' and 'soft'. If the exact value of the emissivity, ε_n, is not known, then for hard coatings or where the type of coating is unknown use the data for $\varepsilon_n = 0.2$, and for soft coatings use the data for $\varepsilon_n = 0.1$. For doors that are half-glazed, the U-value is the average of the non-glazed door and the appropriate U-value for the glazing. For windows and rooflights with metal frames where the thermal break differs from 4 mm, the corrections in Table 4.4 should be applied. Note that if corrections for thermal break *and* rooflight are applicable, both should be made.

Table 4.1 Single glazing U-values for windows, rooflights and doors.

Single glazing description	W/m²K
Windows in wood or PVC-U frames	4.8
Rooflights in dwellings in wood or PVC-U frames	5.1
Rooflights in buildings other than dwellings in wood or PVC-U frames	4.8
Windows in metal frames (4 mm thermal break)	5.7
Solid wooden door	3.0

Table 4.2   Double glazing U-values for windows and rooflights, W/m^{2}K.

Double glazing description	Gap between panes			Adjustment for rooflights in dwellings
	6 mm	12 mm	16 mm or more	
Wood or PVC-U frames				
Air filled	3.1	2.8	2.7	
Low-E, $\varepsilon_n = 0.2$	2.7	2.3	2.1	
Low-E, $\varepsilon_n = 0.15$	2.7	2.2	2.0	
Low-E, $\varepsilon_n = 0.1$	2.6	2.1	1.9	For dwellings only,
Low-E, $\varepsilon_n = 0.05$	2.6	2.0	1.8	add 0.2 for all wood
Argon filled	2.9	2.7	2.6	and PVC-U frames
Low-E, $\varepsilon_n = 0.2$, argon filled	2.5	2.1	2.0	
Low-E, $\varepsilon_n = 0.1$, argon filled	2.3	1.9	1.8	
Low-E, $\varepsilon_n = 0.05$, argon filled	2.3	1.8	1.7	
Metal frames, 4 mm thermal break				
Air filled	3.7	3.4	3.3	
Low-E, $\varepsilon_n = 0.2$	3.3	2.8	2.6	
Low-E, $\varepsilon_n = 0.1$	3.2	2.6	2.5	
Low-E, $\varepsilon_n = 0.05$	3.1	2.5	2.3	For metal frames, see
Argon filled	3.5	3.3	3.2	Table 4.4
Low-E, $\varepsilon_n = 0.2$, argon filled	3.1	2.6	2.5	
Low-E, $\varepsilon_n = 0.1$, argon filled	2.9	2.4	2.3	
Low-E, $\varepsilon_n = 0.05$, argon filled	2.8	2.3	2.1	

Table 4.3   Triple glazing U-values for windows and rooflights, W/m^{2}K.

Triple glazing description	Gap between panes			Adjustment for rooflights in dwellings
	6 mm	12 mm	16 mm or more	
Wood or PVC-U frames				
Air filled	2.4	2.1	2.0	
Low-E, $\varepsilon_n = 0.2$	2.1	1.7	1.6	
Low-E, $\varepsilon_n = 0.1$	2.0	1.6	1.5	For dwellings only,
Low-E, $\varepsilon_n = 0.05$	1.9	1.5	1.4	add 0.2 for all wood
Argon filled	2.2	2.0	1.9	and PVC-U frames
Low-E, $\varepsilon_n = 0.2$, argon filled	1.9	1.6	1.5	
Low-E, $\varepsilon_n = 0.1$, argon filled	1.8	1.4	1.3	
Low-E, $\varepsilon_n = 0.05$, argon filled	1.7	1.4	1.3	
Metal frames, 4 mm thermal break				
Air filled	2.9	2.6	2.5	
Low-E, $\varepsilon_n = 0.2$	2.6	2.2	2.0	
Low-E, $\varepsilon_n = 0.1$	2.5	2.0	1.9	
Low-E, $\varepsilon_n = 0.05$	2.4	1.9	1.8	For metal frames, see
Argon filled	2.8	2.5	2.4	Table 4.4
Low-E, $\varepsilon_n = 0.2$, argon filled	2.4	2.0	1.9	
Low-E, $\varepsilon_n = 0.1$, argon filled	2.2	1.9	1.8	
Low-E, $\varepsilon_n = 0.05$, argon filled	2.2	1.8	1.7	

Table 4.4 Corrections for metal frames with various thermal breaks.

	Correction to U-value, W/m^2K	
Thermal break mm	Window, or rooflight in buildings other than dwellings	Rooflight in dwellings
0 (no break)	+0.3	+0.7
4	0	+0.3
8	−0.1	+0.2
12	−0.2	+0.1
16	−0.2	+0.1

4.1.1 Minimum specifications for windows

Inspection of the U-values in Tables 4.2 and 4.3 reveals the required design specification for a window to meet the maximum U-values of 2.0 or 2.2 W/m^2K given in Tables 2.1 and 3.1. For double glazing with a 6 mm air gap in a wood or UPVC frame, it is not possible to keep within the maximum U-value for any of the listed types of glass. With a 12 mm air gap there are three possibilities, and with a 16 mm air gap there are six. For double glazing in metal frames there is only one possibility, and that will pass only if it includes a 4 mm (or better) thermal break. Extracting these cases from Table 4.2, the double glazed window designs which can be expected to be satisfactory are therefore:

- In wood or UPVC frames:
  - Air filled low-E, $\varepsilon_n = 0.15$, minimum 16 mm air gap
  - Air filled low-E, $\varepsilon_n = 0.10$, minimum 16 mm air gap
  - Air filled low-E, $\varepsilon_n = 0.05$, minimum 12 mm air gap
  - Argon filled low-E, $\varepsilon_n = 0.20$, minimum 16 mm air gap
  - Argon filled low-E, $\varepsilon_n = 0.10$, minimum 12 mm air gap
  - Argon filled low-E, $\varepsilon_n = 0.05$, minimum 12 mm air gap
- In a metal frame with 4 mm thermal break:
  - Argon filled low-E, $\varepsilon_n = 0.05$, minimum 16 mm air gap.

For triple glazing, even though there is a much higher number of glass/frame combinations which satisfy the U-value standard, a significant number do not, especially metal frames with a small air gap.

4.2 Roofs, walls and floors

The tables in AD L1 and AD L2 do not give U-values for complete construc-tions. Rather, they enable the calculation of the minimum thickness of the insulation layer that is necessary to achieve a desired U-value. However, the minimum thickness is applicable only when the insulation layer is perfectly continuous. In practice, U-values may vary because of:

- Air gaps in the insulation
- Mechanical fasteners penetrating the insulation layer
- Precipitation on inverted roofs.

The effect of each of these factors is to add a correction factor to the U-value. Thus, if one or more of these factors is present in a construction, it is necessary to make an appropriate adjustment to the U-value. However, for the purposes of the 'look-up' tables of Appendix A, these correction factors are combined into a single correction factor, ΔU, for which values are given in Table 4.5. If this correction factor is applicable to a particular construction, it is then necessary to select the thickness of insulation to give a lower U-value than that which is desired. Then, when the correction is added on, the net result is the desired U-value. The procedure is therefore first to select the *desired* U-value and decide on the position and method of fixing of the insulation layer, then, using Table 4.5, to obtain ΔU, the correction to the U-value. The *design* U-value is found from the *desired* U-value using:

$$U_{design} = U_{desired} - \Delta U$$

and the tables are used to find the thickness of insulation which will meet the *design* U-value. If this thickness of insulation is used, the desired U-value will be achieved, because:

$$U_{desired} = U_{design} + \Delta U$$

Once the desired U-value has been chosen, and the ΔU correction term (if applicable) has been found, the design U-value may be calculated and the tables used to find the *base* thickness of insulation, defined as the smallest thickness of insulation required to meet a specified U-value *without* allowances for other components in the structure.

   Of course other elements in the construction also provide some thermal resistance, thus allowing a reduction to be made on the base thickness. The true minimum is therefore:

Minimum thickness of insulation = base thickness − allowable reductions

The calculated minimum thickness is unlikely to correspond exactly to any of the thicknesses in which a particular material is supplied, and so the nearest thickness above the minimum must be specified. Nor is it certain that the minimum thickness will be reasonably practical in terms of its installation and fixing. If the result appears impractical, a redesign of the element may be necessary.

Table 4.5 The ΔU correction term for U-values.

Roofs	ΔU, W/m²K
Insulation fixed with nails or screws	0.02
Insulation between joists or rafters	0.01
Insulation between and over joists or rafters	0
Walls	
Timber frame where the insulation partly fills the space between the studs	0.04
Timber frame where the insulation fully fills the space between the studs	0.01
Internal insulation fixed with nails or screws which penetrate the insulation	0.02
External insulation with metal fixings which penetrate the insulation	0.02
Insulated cavity wall with cavity greater than 75 mm and tied with steel vertical-twist ties	0.02
Insulated cavity wall with cavity less than or equal to 75 mm tied with ties other than steel vertical-twist ties	0
Floors	
Suspended timber floor with insulation between joists	0.04
Floor insulation fixed with nails or screws	0.02

4.2.1 Determining the thickness of insulation for roofs

Figure 4.1 illustrates three common roof types, and Table 4.6 gives the base thickness of insulation for these roofs.

Because the base thickness may not be the minimum, the reductions in heat flow due to other elements of the construction must be considered, and further corrections may be made to reduce the base thickness to give the minimum thickness. The amounts by which the base thickness may be reduced are given in Table 4.7.

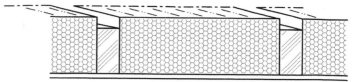

(a) Insulation laid between ceiling joists or rafters

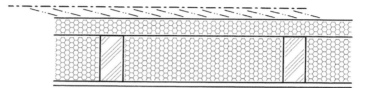

(b) Insulation laid between and over joists or rafters

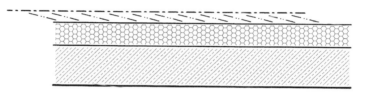

(c) Continuous layer of insulation over a structural base

Fig. 4.1   Three common roof types.

Table 4.6   Base thickness of insulation for roofs.

Design U-value W/m²K	Thermal conductivity of the insulation material, W/mK						
	0.020	0.025	0.030	0.035	0.040	0.045	0.050
	Base thickness of insulation layer, mm						
Insulation laid between ceiling joists or rafters							
0.15	371	464	557	649	742	835	928
0.20	180	224	269	314	359	404	449
0.25	118	148	178	207	237	266	296
0.30	92	110	132	154	176	198	220
0.35	77	91	105	122	140	157	175
0.40	67	78	90	101	116	130	145
Insulation laid between and over joists or rafters							
0.15	161	188	217	247	277	307	338
0.20	128	147	167	188	210	232	255
0.25	108	122	137	153	170	187	205
0.30	92	105	117	130	143	157	172
0.35	77	91	103	113	124	136	148
0.40	67	78	90	101	110	120	130
Continuous layer of insulation							
0.15	131	163	196	228	261	294	326
0.20	97	122	146	170	194	219	243
0.25	77	97	116	135	154	174	193
0.30	64	80	96	112	128	144	160
0.35	54	68	82	95	109	122	136
0.40	47	59	71	83	94	106	118

Table 4.7   Reduction in base thickness of insulation for roof components.

Concrete slab density kg/m³	Thermal conductivity of the insulation material, W/mK						
	0.020	0.025	0.030	0.035	0.040	0.045	0.050
	Reduction (mm) in the base thickness of the insulation, for each 100 mm thickness of the concrete slab						
600	10	13	15	18	20	23	25
800	7	9	11	13	14	16	18
1100	5	6	8	9	10	11	13
1300	4	5	6	7	8	9	10
1700	2	2	3	3	4	4	5
2100	1	2	2	2	3	3	3
Various materials and components	Reduction in the base thickness of the insulation mm						
10 mm plasterboard	1	2	2	2	3	3	3
13 mm plasterboard	2	2	2	3	3	4	4
13 mm sarking board	2	2	3	3	4	4	5
12 mm calcium silicate liner board	1	2	2	2	3	3	4
Roof space (pitched)	4	5	6	7	8	9	10
Roof space (flat)	3	4	5	6	6	7	8
19 mm roof tiles	0	1	1	1	1	1	1
19 mm asphalt (or 3 layers of felt)	1	1	1	1	2	2	2
50 mm screed	2	3	4	4	5	5	6

4.2.1.1   Example calculations for roofs

Example 4.1   Pitched roof with insulation between the joists
Figure 4.2a shows insulation laid between the ceiling joists of a pitched roof which is covered with 19 mm roof tiles. The maximum U-value allowed by the elemental method is, from Table 2.1, 0.16 W/m²K, and it is required to determine the necessary thickness of insulation. First, from Table 4.5, there is a ΔU correction of 0.01 W/m²K. The design (or 'look-up') U-value is therefore:

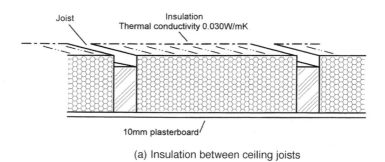

(a) Insulation between ceiling joists

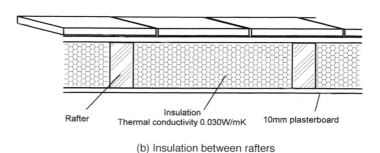

(b) Insulation between rafters

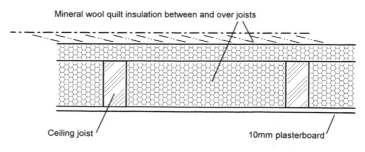

(c) Insulation between and over ceiling joists

Fig. 4.2   Examples 4.1, 4.2 and 4.3.

Insulation between joists $\qquad$ $U_{design} = 0.16 - 0.01 = 0.15 \text{ W/m}^2\text{K}$

Next, use Table 4.6 to find the base thickness of insulation, and Table 4.7 to determine allowable reductions to the base thickness:

From Table 4.6 $\quad$ Base thickness = 557 mm
From Table 4.7 $\quad$ Reduction for $\quad$ 19 mm roof tiles $\qquad$ = 1 mm
$\qquad\qquad\qquad\qquad\qquad\qquad$ Pitched roofspace $\qquad$ = 6 mm
$\qquad\qquad\qquad\qquad\qquad\qquad$ 10 mm plasterboard $\quad$ = 2 mm
$\qquad\qquad\qquad$ Total reduction $\qquad\qquad\qquad\qquad$ = 9 mm

Minimum insulation thickness = base thickness − total reduction = 557 − 9
$$= 548 \text{ mm}$$

In this case, the ceiling joists create a thermal bridge that is not protected by the insulation layer, with the result that a very large thickness of insulation is required. As the joists are likely to be about 100 mm in depth, this thickness of insulation laid between them is not a practicable design solution.

Example 4.2 Pitched roof with insulation between the rafters
Figure 4.2b shows a pitched roof with the insulation laid between the rafters. The maximum U-value allowed by the elemental method is, from Table 2.1, $0.20 \text{ W/m}^2\text{K}$, and it is required to determine the necessary thickness of insulation. First, from Table 4.5, there is a ΔU correction of $0.01 \text{ W/m}^2\text{K}$. The design (or 'look-up') U-value is therefore:

Insulation between rafters $\qquad$ $U_{design} = 0.20 - 0.01 = 0.19 \text{ W/m}^2\text{K}$

Next, use Table 4.6 to find the base thickness of insulation. This requires an interpolation:

At U = 0.19, thickness at λ = 0.030 is $557 - (0.04/0.05) \times (557 - 269)$
$$= 327 \text{ mm}$$
Thus, from Table 4.6 $\quad$ Base thickness = 327 mm

Table 4.7 is now used to determine allowable reductions to the base thickness:

From Table 4.7 $\quad$ Reduction for $\quad$ 19 mm roof tiles $\qquad$ = 1 mm
$\qquad\qquad\qquad\qquad\qquad\qquad$ 10 mm plasterboard $\quad$ = 2 mm
$\qquad\qquad\qquad$ Total reduction $\qquad\qquad\qquad\qquad$ = 3 mm

Minimum insulation thickness = base thickness − total reduction = 327 − 3
$$= 324 \text{ mm}$$

As in the previous example, the bridging effect of the timber has produced a requirement for an insulation layer much thicker than depth of the rafters.

Unless the rafters themselves are also over 300 mm deep, there will be insufficient space between the plasterboard and the tiling for the insulation to be fitted.

Example 4.3   Pitched roof with insulation between and over ceiling joists
Figure 4.2c shows a pitched roof with insulation between and over ceiling joists. The maximum U-value allowed by the elemental method is, from Table 2.1, 0.16 W/m²K, and it is required to determine the necessary thickness of insulation. The insulation material is mineral wool quilt of thermal conductivity (Table 4.16) of 0.042 W/mK. First, from Table 4.5, the ΔU correction is zero. The design (or 'look-up') U-value is therefore:

Insulation between and over joists $U_{design} = 0.16 - 0 = 0.16$ W/m²K

Next, use Table 4.6 to find the base thickness of insulation. This requires a double interpolation:

At U = 0.15, thickness at λ = 0.042 is 277 + (0.002/0.005) × (307 − 277)
= 289 mm
At U = 0.20, thickness at λ = 0.042 is 210 + (0.002/0.005) × (232 − 210)
= 219 mm
At U = 0.16, thickness at λ = 0.042 is 289 − (0.01/0.05) × (289 − 219)
= 275 mm

Thus, from Table 4.6   Base thickness = 275 mm

Table 4.7 is now used to determine allowable reductions to the base thickness:

From Table 4.7   Reduction for   19 mm roof tiles   = 1 mm
                                    Pitched roofspace   = 8 mm
                                    10 mm plasterboard = 3 mm
                    Total reduction   = 12 mm

Minimum insulation thickness = base thickness − total reduction = 275 − 12
= 263 mm

In this case, the thermal bridge created by the ceiling joists is protected by part of the insulation. The resulting thickness of insulation is much less and more practicable than the first two examples, but is still considerable. If access to the roof space is required, then the problem of providing walkways over the insulation without compressing it or penetrating it with fixings will have to be solved.

Example 4.4   Flat roof
Figure 4.3 shows a concrete deck flat roof. The maximum U-value allowed by the elemental method is, from Table 2.1, 0.25 W/m²K, and it is required to determine the necessary thickness of insulation. There are no ΔU corrections, and so:

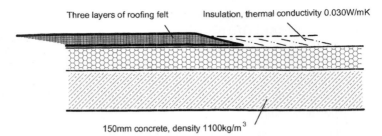

Fig. 4.3   Example 4.4, concrete deck roof.

Continuous layer of insulation $U_{design} = 0.25 - 0 = 0.25 \text{ W/m}^2\text{K}$

Referring to Table 4.6 for a thermal conductivity of 0.030 W/mK, the base thickness of insulation is:

From Table 4.6   Base thickness = 116 mm

There are allowable reductions for the roofing felt and for the concrete deck itself. For 150 mm of concrete of density 1100 kg/m³, the reduction is $1.5 \times 8 =$ 12 mm. Thus:

From Table 4.7   Reduction for   3 layers felt   = 1 mm
                                   150 mm concrete deck   = 12 mm
                    Total reduction   = 13 mm

Minimum insulation thickness = base thickness − total reduction = 116 − 13
                                                                = 103 mm

A solution using this thickness of insulation is feasible but has a number of possible drawbacks. The insulation material itself is unlikely to have any significant compressive strength (otherwise it would not have a thermal conductivity as low as 0.030 W/mK), and so access to the roof deck could not be permitted. Also, because the insulation is immediately beneath the weatherproof membrane, solar radiation on the roof will generate very high surface temperatures, with the possibility of degradation of the membrane due to thermal movement or chemical action.

4.2.2   Determining the thickness of insulation for walls

The procedure is the same as for roofs. Select the desired U-value, decide on the position of the insulation layer and method of fixing, and use Table 4.5 to obtain ΔU, the correction to the U-value. The *design* U-value is found from the *desired* U-value using:

$$U_{design} = U_{desired} - \Delta U$$

Table 4.8 Base thickness of insulation for walls.

Design U-value W/m²K	Thermal conductivity of the insulation material, W/mK						
	0.020	0.025	0.030	0.035	0.040	0.045	0.050
	Base thickness of insulation layer, mm						
0.20	97	121	145	169	193	217	242
0.25	77	96	115	134	153	172	192
0.30	63	79	95	111	127	142	158
0.35	54	67	81	94	107	121	134
0.40	47	58	70	82	93	105	117
0.45	41	51	62	72	82	92	103

Table 4.9 Reduction in base thickness of insulation for wall components.

	Thermal conductivity of the insulation material, W/mK						
	0.020	0.025	0.030	0.035	0.040	0.045	0.050
Concrete blockwork density kg/m³	Reduction (mm) in the base thickness of the insulation, for each 100 mm thickness of the concrete blockwork						
Inner leaf							
600	9	11	13	15	17	20	22
800	7	9	10	12	14	16	17
1000	5	6	8	9	10	11	13
1200	4	5	6	7	8	9	10
1400	3	4	5	6	7	8	8
1600	3	3	4	5	6	6	7
1800	2	2	3	3	4	4	4
2000	2	2	2	3	3	3	4
2400	1	1	2	2	2	2	3
Outer leaf or single leaf wall							
600	8	11	13	15	17	19	21
800	7	9	10	12	14	15	17
1000	5	6	7	8	10	11	12
1200	4	5	6	7	8	9	10
1400	3	4	5	6	6	7	8
1600	3	3	4	5	5	6	7
1800	2	2	3	3	3	4	4
2000	1	2	2	3	3	3	4
2400	1	1	2	2	2	2	3
Various materials and components	Reduction in the base thickness of the insulation mm						
Cavity, 25 mm or more	4	5	5	6	7	8	9
Outer leaf brickwork	3	3	4	5	5	6	6
13 mm plaster	1	1	1	1	1	1	1
13 mm lightweight plaster	2	2	2	3	3	4	4
9.5 mm plasterboard	1	2	2	2	3	3	3
12.5 mm plasterboard	2	2	2	3	3	4	4
Airspace behind plasterboard drylining	2	3	4	4	5	5	6
9 mm sheathing ply	1	2	2	2	3	3	3
20 mm cement render	1	1	1	1	2	2	2
13 mm tile hanging	0	0	0	1	1	1	1

Then use Table 4.8 to find the base thickness of insulation, followed by Table 4.9 for any allowable reduction in the base thickness, and hence obtain the minimum thickness.

For timber framed walls, where the timber frame contains its own integral insulation, there is an additional and often significant reduction in the base thickness of the separate insulation layer. However, this reduction depends on the proportion of the area of the frame that is timber. The Approved Document gives a table for one common case when the proportion of timber is 15% of the wall area, corresponding to 38 mm wide studs at 600 mm centres, with additional timbers at junctions and around openings. This table is given here as Table 4.10. For other area proportions, or frames with insulation material of different thermal conductivity, it is necessary to use an acceptable calculation procedure, such as that given in Appendix B of the Approved Documents (see Chapter 5), to determine either the U-value or the necessary minimum thickness of an insulation layer.

Table 4.10   Reduction in base thickness of insulation for insulated timber frame walls.

Thermal conductivity of insulation within frame W/mK	Thermal conductivity of the insulation material, W/mK						
	0.020	0.025	0.030	0.035	0.040	0.045	0.050
	Reduction (mm) in the base thickness of insulation, for each 100 mm thickness of the timber frame						
0.035	39	49	59	69	79	89	99
0.040	36	45	55	64	73	82	91

4.2.2.1   Example calculations for walls

Example 4.5 Masonry cavity wall
Figure 4.4 shows a masonry cavity wall. The maximum U-value allowed by the elemental method is, from Table 2.1, 0.35 W/m^{2}K, and it is required to determine the necessary thickness of expanded polystyrene board (EPS) insulation, of thermal conductivity 0.040 W/mK. From Table 4.5, there are no ΔU reductions, and so:

$$U_{design} = 0.35 - 0 = 0.35 \text{ W/m}^2\text{K}$$

Now use Tables 4.8 and 4.9 to find the base thickness and the allowable reductions. Note that the reduction for 150 mm of 600 kg/m^3 blockwork is 1.5 × 17 = 25.5 mm. As this is a reduction, it should be rounded to 25 mm.

From Table 4.8	Base thickness = 107 mm	
From Table 4.9	Reduction for brickwork outer leaf	= 5 mm
	cavity	= 7 mm
	concrete blockwork	= 25 mm

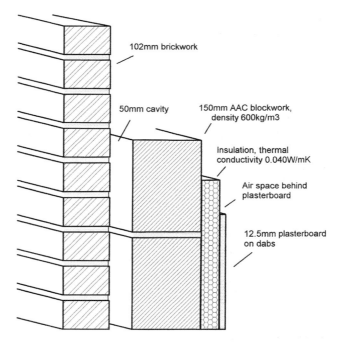

102mm brickwork

50mm cavity

150mm AAC blockwork, density 600kg/m3

Insulation, thermal conductivity 0.040W/mK

Air space behind plasterboard

12.5mm plasterboard on dabs

Fig. 4.4   Masonry cavity wall with internal insulation.

$$\text{air space behind plasterboard} = 5 \text{ mm}$$
$$\text{plasterboard} = 3 \text{ mm}$$
$$\text{Total reduction} = 45 \text{ mm}$$

$$\text{Minimum insulation thickness} = \text{base thickness} - \text{total reduction} = 107 - 45$$
$$= 61 \text{ mm}$$

The next available thickness of EPS board above the 62 mm minimum is likely to be 75 mm. This should have sufficient rigidity to be suitable for fixing, as shown in Fig. 4.4.

Example 4.6   Masonry wall with cavity fill
Figure 4.5 shows a masonry wall with the cavity completely filled with poly-urethane foam insulation of thermal conductivity 0.040 W/mK. The brickwork and blockwork are tied with stainless steel vertical-twist ties. The maximum U-value allowed by the elemental method is, from Table 2.1, 0.35 W/m²K, and it is required to determine the necessary thickness of insulation. From Table 4.5, there is a ΔU reduction of 0.02 W/m²K for the wall ties, and so:

$$U_{design} = 0.35 - 0.02 = 0.33 \text{ W/m}^2\text{K}$$

Next, use Table 4.8 to find the base thickness of insulation. This requires an interpolation:

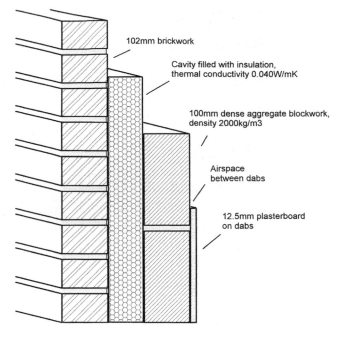

102mm brickwork

Cavity filled with insulation, thermal conductivity 0.040W/mK

100mm dense aggregate blockwork, density 2000kg/m3

Airspace between dabs

12.5mm plasterboard on dabs

Fig. 4.5   Masonry wall with cavity fill.

At U = 0.33, thickness at λ = 0.040 is $127 - (0.03/0.05) \times (127 - 107)$
$$= 115 \, \text{mm}$$

Thus, from Table 4.8   Base thickness = 115 mm

Now use Table 4.9 to obtain allowable reductions:

From Table 4.9   Reduction for brickwork outer leaf   =   5 mm
                                    concrete blockwork   =   3 mm
                                    air space behind plasterboard =   5 mm
                                    plasterboard   =   3 mm
                     Total reduction   = 16 mm

Minimum insulation thickness = base thickness − total reduction = 115 − 16
$$= 99 \, \text{mm}$$

This shows that the wall will have to be constructed with a 100 mm cavity. If this is not acceptable, an alternative solution would be to use concrete blocks with a lower density and/or a greater thickness. The calculation would have to be repeated to see if the reduction in the minimum insulation thickness is sufficient.

Example 4.7   Masonry cavity wall with partial cavity fill
Figure 4.6 shows a masonry wall with the cavity partially filled with insulation. The brickwork and blockwork are tied with stainless steel vertical-twist ties. The

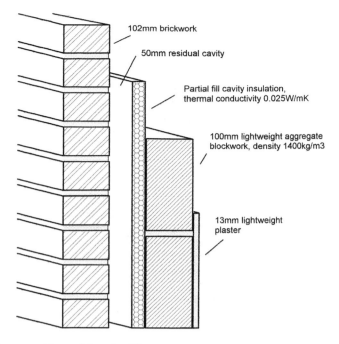

102mm brickwork

50mm residual cavity

Partial fill cavity insulation,
thermal conductivity 0.025W/mK

100mm lightweight aggregate
blockwork, density 1400kg/m3

13mm lightweight
plaster

Fig. 4.6   Masonry with partial cavity fill.

maximum U-value allowed by the elemental method is, from Table 2.1, 0.35 W/ m^{2}K, and it is required to determine the necessary thickness of insulation. From Table 4.5, there is a ΔU reduction of 0.02 W/m^{2}K for the wall ties, and so:

$$U_{design} = 0.35 - 0.02 = 0.33 \text{ W/m}^2\text{K}$$

Next, use Table 4.8 to find the base thickness of insulation. This requires an interpolation:

At U = 0.33, thickness at λ = 0.025 is $79 - (0.03/0.05) \times (79 - 67)$
$$= 72 \text{ mm}$$
Thus, from Table 4.8   Base thickness = 72 mm

Now use Table 4.9 to obtain allowable reductions:

From Table 4.9   Reduction for   brickwork outer leaf   =   3 mm

cavity   =   5 mm

concrete blockwork   =   4 mm

lightweight plaster   =   2 mm

Total reduction   = 14 mm

Minimum insulation thickness = base thickness − total reduction = 72 − 14
$$= 58 \text{ mm}$$

The minimum insulation thickness is much less than in the previous example. Nevertheless, if a 50 mm residual cavity is to be preserved, the overall cavity width will have to be of the order of 100 mm.

Example 4.8   Timber frame wall
Figure 4.7 shows a timber frame wall. The maximum U-value allowed by the elemental method is, from Table 2.1, 0.35 W/m²K. The 90 mm timber frame has insulation that fully fills the space between the studs. It is required to determine the necessary thickness of a continuous insulation layer of EPS board that is to be fixed between the sheathing ply and the timber frame. From Table 4.5, there is a ΔU reduction of 0.01 for a timber frame where the insulation fully fills the space between the studs. However, the insulation layer which is being added is continuous, and so the ΔU reduction may be taken as zero.

$$U_{design} = 0.35 - 0 = 0.35 \text{ W/m}^2\text{K}$$

Now use Table 4.8 to find the base thickness:

From Table 4.8   Base thickness = 107 mm

Now use Tables 4.9 and 4.10 to obtain allowable reductions:

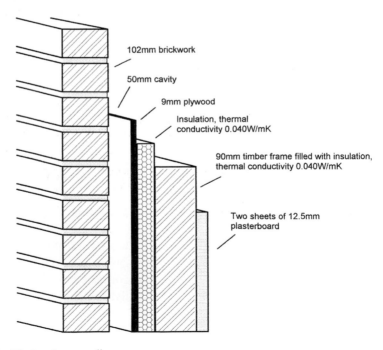

102mm brickwork

50mm cavity

9mm plywood

Insulation, thermal conductivity 0.040W/mK

90mm timber frame filled with insulation, thermal conductivity 0.040W/mK

Two sheets of 12.5mm plasterboard

Fig. 4.7   Timber frame wall.

From Table 4.9   Reduction for   brickwork outer leaf   = 5 mm
                                   cavity   = 7 mm
                                   sheathing ply   = 3 mm
                                   2 sheets plasterboard   = 6 mm
                   Reductions from Table 4.9   = 21 mm
From Table 4.10   Reduction for timber frame (73 × 90/100)   = 66 mm
                   Total reduction   = 87 mm

Minimum insulation thickness = base thickness − total reduction = 107 − 87
                                                                = 20 mm

This thickness of insulation should provide a suitable solution.

4.2.3   Determining the thickness of insulation for floors

Ground floors must be considered separately from upper floors. In the case of ground floors, the U-value depends not only on the insulation but also on the size and shape of the floor. Both size and shape are taken into account by means of the ratio P/A, where P is the perimeter length in metres of the whole of the ground floor, and A is the total area in square metres of the ground floor. Further, AD L1 does not give any allowable reductions for components in the floor structure, and so the 'base' thickness of insulation is also the minimum thickness. Data is given for the three principal types of floor, i.e. solid floors in contact with the ground (Fig. 4.8 and Table 4.11), suspended timber ground floors (Fig. 4.9 and Table 4.12) and suspended concrete beam and block floors (Fig. 4.10 and Table 4.13).

4.2.3.1   *Solid floor in contact with the ground*

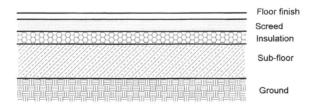

Fig. 4.8   Solid floor in contact with the ground.

Table 4.11   Thickness of insulation for solid floors in contact with the ground.

Design U-value W/m²K	P/A ratio m/m²	Thermal conductivity of the insulation material, W/mK						
		0.020	0.025	0.030	0.035	0.040	0.045	0.050
		Base (minimum) thickness of insulation layer, mm						
0.020	1.00	81	101	121	142	162	182	202
	0.90	80	100	120	140	160	180	200
	0.80	78	98	118	137	157	177	196
	0.70	77	96	115	134	153	173	192
	0.60	74	93	112	130	149	167	186
	0.50	71	89	107	125	143	160	178
	0.40	67	84	100	117	134	150	167
	0.30	60	74	89	104	119	134	149
	0.20	46	57	69	80	92	103	115
0.25	1.00	61	76	91	107	122	137	152
	0.90	60	75	90	105	120	135	150
	0.80	58	73	88	102	117	132	146
	0.70	57	71	85	99	113	128	142
	0.60	54	68	82	95	109	122	136
	0.50	51	64	77	90	103	115	128
	0.40	47	59	70	82	94	105	117
	0.30	40	49	59	69	79	89	99
	0.20	26	32	39	45	52	58	65
0.30	1.00	48	60	71	83	95	107	119
	0.90	47	58	70	81	93	105	116
	0.80	45	56	68	79	90	102	113
	0.70	43	54	65	76	87	98	108
	0.60	41	51	62	72	82	92	103
	0.50	38	47	57	66	76	85	95
	0.40	33	42	50	59	67	75	84
	0.30	26	33	39	46	53	59	66
	0.20	13	16	19	22	25	28	32

4.2.3.2   *Suspended timber ground floor*

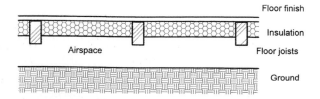

Fig. 4.9   Suspended timber ground floor.

Table 4.12 Thickness of insulation for suspended timber ground floors.

Design U-value W/m²K	P/A ratio m/m²	Thermal conductivity of the insulation material, W/mK						
		0.020	0.025	0.030	0.035	0.040	0.045	0.050
		Base (minimum) thickness of insulation layer, mm						
0.020	1.00	127	145	164	182	200	218	236
	0.90	125	144	162	180	198	216	234
	0.80	123	142	160	178	195	213	230
	0.70	121	139	157	175	192	209	226
	0.60	118	136	153	171	188	204	221
	0.50	114	131	148	165	181	198	214
	0.40	109	125	141	157	173	188	204
	0.30	99	115	129	144	159	173	187
	0.20	82	95	107	120	132	144	156
0.25	1.00	93	107	121	135	149	162	176
	0.90	92	106	119	133	146	160	173
	0.80	90	104	117	131	144	157	170
	0.70	88	101	114	127	140	153	166
	0.60	85	98	111	123	136	148	161
	0.50	81	93	106	118	130	142	154
	0.40	75	87	99	110	121	132	143
	0.30	66	77	87	97	107	117	127
	0.20	49	57	65	73	81	88	96
0.30	1.00	71	82	93	104	114	125	135
	0.90	70	80	91	102	112	122	133
	0.80	68	78	89	99	109	119	129
	0.70	66	76	86	96	106	116	126
	0.60	63	73	82	92	102	111	120
	0.50	59	68	78	87	96	104	113
	0.40	53	62	70	79	87	95	103
	0.30	45	52	59	66	73	80	87
	0.20	28	33	38	42	47	51	56

4.2.3.3 *Suspended concrete ground floor*

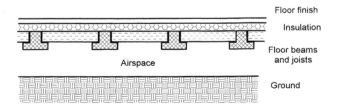

Fig. 4.10 Suspended concrete ground floor.

Table 4.13 Thickness of insulation for suspended concrete beam and block ground floors.

Design U-value W/m²K	P/A ratio m/m²	Thermal conductivity of the insulation material, W/mK						
		0.020	0.025	0.030	0.035	0.040	0.045	0.050
		Base (minimum) thickness of insulation layer, mm						
0.20	1.00	82	103	123	144	164	185	205
	0.90	81	101	122	142	162	183	203
	0.80	80	100	120	140	160	180	200
	0.70	79	99	118	138	158	177	197
	0.60	77	96	116	135	154	173	193
	0.50	75	93	112	131	150	168	187
	0.40	71	89	107	125	143	161	178
	0.30	66	82	99	115	132	148	165
	0.20	56	69	83	97	111	125	139
0.25	1.00	62	78	93	109	124	140	155
	0.90	61	76	92	107	122	138	153
	0.80	60	75	90	105	120	135	150
	0.70	59	74	88	103	118	132	147
	0.60	57	71	86	100	114	128	143
	0.50	55	68	82	96	110	123	137
	0.40	51	64	77	90	103	116	128
	0.30	46	57	69	80	92	103	115
	0.20	36	45	54	62	71	80	89
0.30	1.00	49	61	73	85	97	110	122
	0.90	48	60	72	84	96	108	120
	0.80	47	59	70	82	94	105	117
	0.70	45	57	68	80	91	102	114
	0.60	44	55	66	77	88	98	109
	0.50	41	52	62	72	83	93	104
	0.40	38	48	57	67	76	86	95
	0.30	33	41	49	57	65	73	81
	0.20	22	28	33	39	44	50	56

4.2.3.4   Determining the thickness of insulation for upper floors

For upper floors, i.e. floors above an external space, Table 4.14 applies, together with the allowable reductions in Table 4.15. Note that in Table 4.14 it is assumed that the proportion by area of structural timber in the timber floor construction is 12%, corresponding to 48 mm wide timber joists at 400 mm centres. The U-value for other proportions of timber must be calculated using the procedures in Chapter 5.

Table 4.14   Thickness of insulation for upper floors of timber construction.

	Design U-value W/m²K	Thermal conductivity of the insulation material, W/mK						
		0.020	0.025	0.030	0.035	0.040	0.045	0.050
		Base thickness of insulation layer, mm						
Timber	0.20	167	211	256	298	341	383	426
construction	0.25	109	136	163	193	225	253	281
	0.30	80	100	120	140	160	184	208
Concrete	0.20	95	119	142	166	190	214	237
construction	0.25	75	94	112	131	150	169	187
	0.30	62	77	92	108	123	139	154

Table 4.15   Reduction in base thickness for upper floor components.

	Thermal conductivity of the insulation material, W/mK						
	0.020	0.025	0.030	0.035	0.040	0.045	0.050
Component	Reduction in the base thickness of the insulation, mm						
10 mm plasterboard	1	2	2	2	3	3	3
19 mm timber flooring	3	3	4	5	5	6	7
50 mm screed	2	3	4	4	5	5	6

4.2.3.5   Example calculations for floors

Example 4.9   Solid floor in contact with the ground
Figure 4.11 shows a solid floor. The maximum U-value allowed by the elemental method, from Tables 2.1 and 3.1, is 0.25 W/m²K, and it is desired to find the necessary thickness of insulation of thermal conductivity 0.025 W/mK. Table 4.11 must be used, and this requires the perimeter to area ratio to be calculated:

Floor perimeter $= 6 + 2 + 4 + 4 + 10 + 6 = 32\,m$
Floor area $= (6 \times 6) + (4 \times 4) = 52\,m^2$
Perimeter to area ratio, P/A: $\dfrac{P}{A} = \dfrac{32}{52} = 0.615 \cong 0.6$

From Table 4.11, the necessary minimum thickness of insulation is 68 mm.

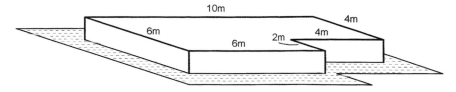

Fig. 4.11   Solid floor.

Example 4.10   Suspended timber ground floor
If the floor shown in Fig. 4.11 was a suspended timber floor of the same perimeter shape and dimensions, the perimeter to area ratio would still be 0.6, but it would be necessary to use Table 4.12. To achieve the same U-value of 0.25 W/m²K, and using insulation of thermal conductivity 0.025 W/mK, the necessary minimum thickness of insulation placed between the joists is 98 mm.

Example 4.11   Upper floor, timber construction
Figure 4.12 shows an upper floor in timber construction. The joists are 45 mm wide and are at 400 mm centres. The maximum U-value allowed by the elemental method is, from Tables 2.1 and 3.1, 0.25 W/m²K, and it is desired to find the necessary thickness of insulation of thermal conductivity 0.030W/mK. The proportion by area of structural timber in the floor is $(45/400) \times 100 = 11.25\%$. As this is less than 12%, Table 4.14 will very slightly overestimate the insulation thickness and therefore can still be used. If the proportion of structural timber had been greater than 12%, Table 4.14 would be inappropriate and a suitable calculation method [17, 18] would have to be used instead. Using Tables 4.14 and 4.15:

From Table 4.14   Base thickness = 163 mm
From Table 4.15   Reduction for 10 mm plasterboard   = 2 mm
                                 19 mm floorboards   = 4 mm
              Total reductions   = 6 mm

Minimum insulation thickness = base thickness − total reduction = 163 − 6
                                                                = 157 mm

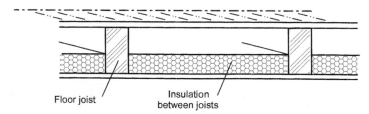

Fig. 4.12   Upper floor, timber construction.

Floor joists are typically not less than 175 mm deep, and so this thickness of insulation can be accommodated.

Example 4.12   Upper floor, concrete
Figure 4.13 shows an upper floor in concrete construction. The required U-value is 0.25 W/m²K, and insulation of thermal conductivity 0.030 W/mK is to be used.

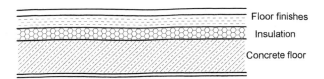

Fig. 4.13 Upper floor – concrete.

From Table 4.14   Base thickness = 112 mm
From Table 4.15   Reduction for 50 mm screed   = 4 mm
                   Total reduction   = 4 mm

Minimum insulation thickness = base thickness − total reduction = 112 − 4
                                                                = 108 mm

The stability and compressive strength of this thickness of insulation immediately beneath the floor finishes must be considered.

4.3   Thermal conductivity and density of building materials

Table 4.16 lists the thermal conductivities and densities of some common building materials.

Table 4.16 Thermal conductivity and density of common building materials.

	Density Kg/m^3	Thermal conductivity W/mK
Walls		
Brickwork (outer leaf)	1700	0.77
Brickwork (inner leaf)	1700	0.56
Lightweight aggregate concrete block	1400	0.57
Autoclaved aerated concrete block	600	0.18
Concrete, medium density (inner leaf)	1800	1.13
	2000	1.33
	2200	1.59
Concrete, high density	2400	1.93
Reinforced concrete, 1% steel	2300	2.30
Reinforced concrete, 2% steel	2400	2.50
Mortar, protected	1750	0.88
Mortar, exposed	1750	0.94
Gypsum	600	0.18
	900	0.30
	1200	0.43
Gypsum plasterboard	900	0.25
Sandstone	2600	2.30
Limestone, soft	1800	1.10
Limestone, hard	2200	1.70
Fibreboard	400	0.10
		Contd

Table 4.16 *Contd*

	Density Kg/m^3	Thermal conductivity W/mK
Plasterboard	900	0.25
Tiles, ceramic	2300	1.30
Timber, softwood	500	0.13
Timber, hardwood	700	0.18
Timber, plywood and chipboard	500	0.13
Wall ties, stainless steel	7900	17.00
Surface finishes		
External rendering	1300	0.57
Plaster, dense	1300	0.57
Plaster, lightweight	600	0.18
Roofs		
Aerated concrete slab	500	0.16
Asphalt	2100	0.70
Felt/bitumen layers	1100	0.23
Screed	1200	0.41
Stone chippings	2000	2.00
Tiles, clay	2000	1.00
Tiles, concrete	2100	1.50
Wood wool slab	500	0.10
Floors		
Cast concrete	2000	1.35
Metal tray, steel	7800	50.00
Screed	1200	0.41
Timber, softwood	500	0.13
Timber, hardwood	700	0.18
Timber, plywood and chipboard	500	0.13
Insulation		
Expanded polystyrene (EPS) board	15	0.040
Mineral wool quilt	12	0.042
Mineral wool batt	25	0.038
Phenolic foam board	30	0.025
Polyurethane board	30	0.025

5 The Calculation of U-values for Walls

Appendix B of the Approved Documents deals with the calculation of U-values from the thickness and thermal conductivity of the materials that make up a construction element. The methods described in Appendix B are not comprehensive but are adequate for many typical constructions. More sophisticated methods are necessary if the construction is complex, particularly if there is significant thermal bridging or if there is two or three dimensional heat flow at corners or reveals.

5.1 Background theory

The lower the U-value of a construction element, the more significant is the effect of thermal bridging on the calculation of the U-value. Consequently it is usually necessary to include thermal bridging in the calculation method. The theory is based on the calculation of thermal resistances. For a single layer of material, the thermal resistance R is given by:

$$R = \frac{d}{\lambda}$$

where d is the thickness of the layer in metres, and λ is the thermal conductivity. The combined resistance of several materials depends on whether the heat flows through them sequentially, i.e. in series or in parallel, as shown in Figure 5.1.

Resistances in series
For three materials in series (Fig. 5.1a), the total combined resistance R_T is given by:

$$R_T = R_1 + R_2 + R_3$$

Resistances in parallel
For three materials in parallel (Fig. 5.1b), the total combined resistance R_T is given by:

$$\frac{1}{R_T} = \frac{F_1}{R_1} + \frac{F_2}{R_2} + \frac{F_3}{R_3}$$

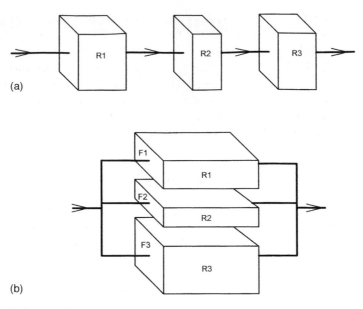

(a)

(b)

Fig. 5.1   Resistances in series (a) and in parallel (b).

where F_1, F_2 and F_3 are the cross-sectional areas of each material expressed as a fraction of the total.

Once the total resistance of a structure has been found, the U-value is its reciprocal:

$$U = \frac{1}{R_T}$$

Construction elements with materials in series and in parallel

Many practical construction elements consist of several layers through which heat passes in series, with some components embedded within them through which the heat passes in parallel (the thermal bridges). The total thermal resistance can be calculated in several ways, and the Approved Documents present one of the simplest and most direct. The method is suitable when the bridging material is timber or mortar, or some other material which is thermally similar. It is not suitable when the bridging material is metal, nor is it suitable for ground floors and basements. The method calculates an upper resistance limit for the construction element, R_{upper}, and a lower resistance limit, R_{lower}, and then finds R_T by taking the average:

$$R_T = \frac{1}{2} \left(R_{upper} + R_{lower} \right)$$

The upper resistance limit is found by taking each possible heat flow path separately and calculating its resistance as the sum of the resistances of each component taken in series. When all paths have been calculated, their resistances are combined in parallel, taking account of their relative areas to give R_{upper}. The lower resistance limit is found by first converting each thermal bridge in the construction into a single equivalent resistance by combining the elements of the bridge in parallel. When all bridges have been converted, the construction should consist of a single heat flow path whose resistance is found by summing its components in series to give R_{lower}. The example calculations which follow should clarify the method.

5.1.1 Corrections to calculated U-values

The ΔU correction factors described in Appendix A and tabulated in Table 4.5 can be treated more rigorously when calculating U-values (see Appendix D of BS EN 150 6946 [12] for details). The procedure is:

$$\text{Corrected U-value} \quad U_c = U + \Delta U$$
$$\Delta U = \Delta U_g + \Delta U_f + \Delta U_r$$
$$\Delta U_g = \text{correction for air gaps}$$
$$\Delta U_f = \text{correction for mechanical fasteners}$$
$$\Delta U_r = \text{correction for inverted roofs}$$

5.1.1.1 Correction for air gaps

$$\Delta U_g = \Delta U^{11} \times \left(\frac{R_1}{R_T}\right)^2$$

where R_1 is thermal resistance of layer containing the gaps
R_T is total thermal resistance of the whole component
ΔU^{11} is obtained from Table 5.1.

Table 5.1 Correction factors for air gaps.

Level	ΔU^{11},W/m^{2}K	Type of air gap
0	0.00	Insulation installed in such a way that no air circulation is possible on the warm side of the insulation. No air gaps penetrating the entire insulation layer.
1	0.01	Insulation installed in such a way that no air circulation is possible on the warm side of the insulation. Air gaps may penetrate the insulation layer.
2	0.04	Air circulation is possible on the warm side of the insulation. Air gaps may penetrate the insulation layer.

Reproduced with permission from BS EN ISO 6946

5.1.1.2   Correction for mechanical fasteners

$$\Delta U_f = \alpha \lambda_f n_f A_f$$

where λ_f is thermal conductivity of the fastener
n_f is number of fasteners per square metre
A_f is cross-sectional area of one fastener
α is obtained from Table 5.2.

Table 5.2   Corrections for mechanical fasterners.

Type of fastener	α, m^{-1}
Wall tie between masonry leaves	6
Roof fixing	5

Reproduced with permission from BS EN ISO 6946

Corrections for fasteners must *not* be applied when:

- The wall ties are across an empty cavity
- The wall ties are between a masonry leaf and timber studs
- The thermal conductivity of the fastener, or part of it, is less than 1 $Wm^{-1}K$.

5.1.1.3   Correction for inverted roofs

The correction for an inverted roof is not yet available.

5.1.2   The U-value via an unheated space

The precise calculation of the heat flow through a building element, and then via an unheated space to the outside, requires complex procedures. These can be found in BS EN ISO 13789 [10]. However, for the purposes of Part L, a simpler procedure in which the unheated space is assumed to behave like an additional homogeneous plane layer is usually adequate. With this assumption, the extra thermal resistance of an unheated space may be included in the calculation of the U-value of an element using the formula:

$$U = \frac{1}{R_o + R_{extra}}$$

where   U is the U-value of the element including the effect of the unheated space
R_0 is the thermal resistance of the element as if exposed directly to the outside
R_{extra} is the extra thermal resistance due to the unheated space.

This formula is acceptable provided R_{extra} is small compared to R_0, say if $R_{extra} < 0.3R_0$. Values of R_{extra} are given in SAP 2001 [19] for some typical unheated spaces attached to dwellings, including:

- Single and double garages in various configurations
- Stairwells
- Access corridors
- Conservatories
- Roof spaces adjacent to a room in a roof.

These values, which are shown in Tables 5.3 and 5.4, can also be applied to similar situations in other buildings. For other unheated spaces it may be possible to calculate R_{extra} from:

$$R_{extra} = 0.09 + 0.4(A_{INT}/A_{EXT})$$

where A_{INT} is the total area of the elements separating the internal heated space from the unheated space, and A_{EXT} is the total area of the elements separating the unheated space from the outside. However, if this formula yields a result for R_{extra} greater than $0.5\,W/m^2K$, it should not be used.

Table 5.3   Extra thermal resistance due to unheated spaces – garages.

Garage type and description	R_{extra} m^{2}K/W	
	Garage inside insulation layer of building	Garage outside insulation layer of building
Single, fully integral, sharing side wall, end wall and floor with building	0.68	0.33
Single, fully integral, sharing side wall and floor with building	0.54	0.25
Single, partially integral, projecting forward, sharing part of side wall, part of floor and end wall with building	0.56	0.26
Single, adjacent, sharing side wall only with building	0.23	0.09
Double, fully integral, sharing side wall, end wall and floor with building	0.59	0.28
Double, half integral, sharing side wall, half of end wall and half of floor with building	0.34	0.17
Double, partially integral, projecting forward, sharing part of side wall, part of floor and end wall with building	0.28	0.13
Double, adjacent, sharing side wall only with building	0.13	0.05

Adapted with permission from SAP 2001 [19]

Table 5.4 Extra thermal resistance due to unheated spaces – various.

Type and description of unheated space	R_{extra} m²K/W
Stairwell between heated space and external (exposed) wall	0.82
Stairwell between heated space and internal (not exposed) wall	0.90
Access corridor between heated space and external (exposed) wall, with another corridor above *or* below	0.31
Access corridor between heated space and external (exposed) wall, with another corridor above *and* below	0.28
Access corridor between heated space and internal (not exposed) wall, with another corridor above *or* below	0.43
Conservatory, double glazed, sharing one wall with heated space	0.06
Conservatory, double glazed, sharing two walls with heated space, i.e. in the angle between two walls	0.14
Conservatory, double glazed, sharing three walls with heated space, i.e. recessed	0.25
Conservatory, single glazed, sharing any number of walls with heated space	0.10
Loft space, between the roof covering and the wall of a heated room formed within a pitched roof above an insulated ceiling, for heat flow horizontally through the wall of the heated room	0.50
Loft space, between the roof covering and the wall of a heated room formed within a pitched roof above an insulated ceiling, for heat flow vertically from/through the ceiling of the room below	0.50

Adapted with permission from SAP 2001 [19]

5.2 Example calculations

The calculation method is most conveniently explained by means of examples.

Example 5.1 Cavity wall
Figure 5.2 shows a cavity wall consisting of external brickwork, cavity, light-weight blockwork, mineral wool insulation within a timber sub-frame, and internal plasterboard. The blockwork and the mineral wool are the main providers of thermal insulation in this construction, and both suffer from thermal bridging. The blockwork is bridged by the mortar joints, and the mineral wool is bridged by the timber frame. In each case, the proportion of the area bridged is:

    Blockwork 93% of area   mortar joints 7% of area
    Mineral wool 88% of area   timber battens 12% of area.

Table 5.5 gives the thermal data for the wall.

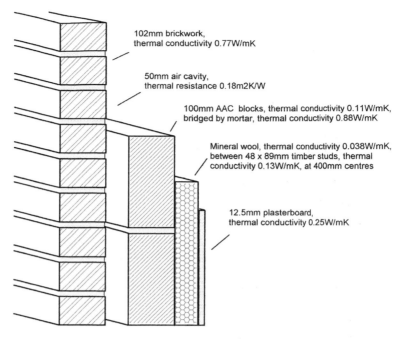

102mm brickwork, thermal conductivity 0.77W/mK

50mm air cavity, thermal resistance 0.18m2K/W

100mm AAC blocks, thermal conductivity 0.11W/mK, bridged by mortar, thermal conductivity 0.88W/mK

Mineral wool, thermal conductivity 0.038W/mK, between 48 x 89mm timber studs, thermal conductivity 0.13W/mK, at 400mm centres

12.5mm plasterboard, thermal conductivity 0.25W/mK

Fig. 5.2   Brick and blockwork cavity wall.

Table 5.5   Thermal data for cavity wall.

Material	Thickness mm	Thermal conductivity W/mK	Thermal resistance m^2K/W
External surface	—	—	0.040
Outer brickwork	102	0.77	—
Cavity, unvented	—	—	0.180
AAC blocks	100	0.11	0.909
Mortar	100	0.88	0.114
Mineral wool insulation	89	0.038	2.342
Timber battens	89	0.13	0.685
Plasterboard	12.5	0.25	0.050
Internal surface	—	—	0.130

The upper resistance limit, R_{upper}

Each possible heat flow path through the wall is considered separately, and in this case it can be seen that there are four such paths, as shown in Fig. 5.3. The resistance of each path is calculated on the basis that the materials are in series, and then the four paths are combined on the basis that they are in parallel. The first part of the calculation is illustrated in Table 5.6. The four paths are then combined in parallel to find R_{upper}:

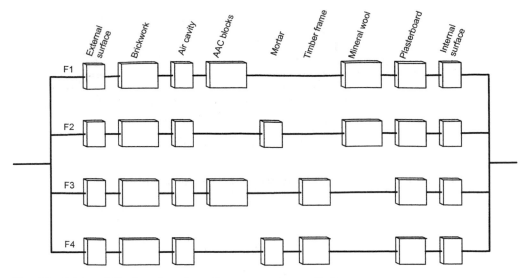

Fig. 5.3   Brick and blockwork cavity wall – upper resistance limit.

Table 5.6   Calculation of the upper resistance limit, cavity wall.

| | Thermal resistance, m²K/W | | | |
	Path 1	Path 2	Path 3	Path 4
External surface resistance	0.040	0.040	0.040	0.040
Resistance of brickwork	0.132	0.132	0.132	0.132
Resistance of cavity	0.180	0.180	0.180	0.180
Resistance of AAC blocks	0.909	—	0.909	—
Resistance of mortar	—	0.114	—	0.114
Resistance of mineral wool	2.342	2.342	—	—
Resistance of timber	—	—	0.685	0.685
Resistance of plasterboard	0.050	0.050	0.050	0.050
Internal surface resistance	0.130	0.130	0.130	0.130
Total thermal resistance of path	3.783	2.988	2.126	1.331
Fractional area of path	93% × 88% = 0.818	7% × 88% = 0.062	93% × 12% = 0.112	7% × 12% = 0.008

$$\frac{1}{R_{upper}} = \frac{F_1}{R_1} + \frac{F_2}{R_2} + \frac{F_3}{R_3} + \frac{F_4}{R_4} = \frac{0.818}{3.783} + \frac{0.062}{2.988} + \frac{0.112}{2.126} + \frac{0.008}{1.331} = 0.2957$$

$$R_{upper} = 3.382 \, m^2K/W$$

The lower resistance limit, R_{lower}
Each thermal bridge in the construction element is first converted to a single combined resistance, as shown in Fig. 5.4. Using these combined resistances, the construction can then be considered as a single heat flow path with all components in series. Thus in the present example, the AAC blocks and the mortar form one thermal bridge, and their combined resistance, R_{bm}, is found from:

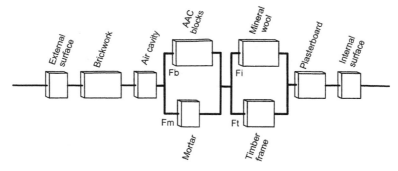

Fig. 5.4 Brick and blockwork cavity wall – lower resistance limit.

$$\frac{1}{R_{bm}} = \frac{F_{blocks}}{R_{blocks}} + \frac{F_{mortar}}{R_{mortar}} = \frac{0.93}{0.909} + \frac{0.07}{0.114} = 1.637$$

$$R_{bm} = 0.611\,\text{m}^2\text{K/W}$$

The mineral wool insulation in its timber frame form another thermal bridge, and their combined resistance, R_{it}, is found from:

$$\frac{1}{R_{it}} = \frac{F_{insulation}}{R_{insulation}} + \frac{F_{timber}}{R_{timber}} = \frac{0.88}{2.342} + \frac{0.12}{0.685} = 0.5509$$

$$R_{it} = 1.815\,\text{m}^2\text{K/W}$$

The combined resistances may now be used with the other resistances in the chain to find the lower resistance limit, as shown in Table 5.7.

Note that R_{upper} is an overestimate of the true resistance, whereas R_{lower} is an

Table 5.7 Calculation of the lower resistance limit, cavity wall.

	Thermal resistances, m²K/W Thermal bridges		
	Components	**Combined**	
External surface resistance	—	—	0.040
Resistance of brickwork	—	—	0.132
Resistance of cavity	—	—	0.180
Resistance of AAC blocks (93%)	0.909 ⎱	0.611	0.611
Resistance of mortar (7%)	0.114 ⎰		
Resistance of mineral wool (88%)	2.342 ⎱	1.815	1.815
Resistance of timber (12%)	0.685 ⎰		
Resistance of plasterboard	—	—	0.050
Internal surface resistance	—	—	0.130
Total thermal resistance, R_{lower}			2.958

underestimate. The average of these is very close to the true value. Hence, the total resistance of the wall is found from

$$R_T = \frac{1}{2}\left(R_{upper} + R_{lower}\right) = \frac{1}{2}(3.382 + 2.958) = 3.170 \text{ m}^2\text{K/W}$$

and the U-value is

$$U = \frac{1}{R_T} = \frac{1}{3.170} = 0.315 \text{ W/m}^2\text{K}$$

Corrections to the U-value for air gaps and mechanical fixings
If there are small air gaps or mechanical fixings (such as wall ties) penetrating the insulation layer, it may be necessary to add a correction, ΔU_g , to the U-value. The correction is required if ΔU_g is 3% or more of the uncorrected U-value, but may be ignored if it is less than 3%. The correction is calculated from:

$$\Delta U_g = \Delta U^{11} \times \left(\frac{R_1}{R_T}\right)^2$$

In this wall, the fixing of the mineral wool insulation in its timber sub-frame is such that there is no air movement on the warm side, but there are some air gaps penetrating the insulation layer. As the air gaps are in the mineral wool and timber sub-frame, $R_1 = R_{it} = 1.815$. Referring to Table 5.1, the correction for air gaps is level 1, and so $\Delta U^{11} = 0.01$. With $R_T = 3.170$, the correction is thus:

$$\Delta U_g = 0.01 \times \left(\frac{1.815}{3.170}\right)^2 = 0.003 \text{ W/m}^2\text{K}$$

As this is less than 3% of U, it may be ignored. The final U-value is rounded to two decimal places:

$$U = 0.32 \text{ W/m}^2\text{K}$$

Example 5.2   Timber framed wall
Figure 5.5 shows a timber framed wall consisting of an outer layer of brickwork, a clear ventilated cavity, 10 mm plywood, 38 × 140 mm timber stud framing with 140 mm mineral wool quilt insulation placed between the studs, and two sheets of 12.5 mm plasterboard with an integral vapour check. The timber studs account for 15% of the area, corresponding to 38 mm studs at 600 mm centres, with allowances for horizontal noggins and additional framing at junctions and around openings. The thermal data for this wall is given in Table 5.8.

The upper resistance limit, R_{upper}
Each possible heat flow path through the wall is considered separately, and in this case it can be seen that there are two such paths. This is illustrated in

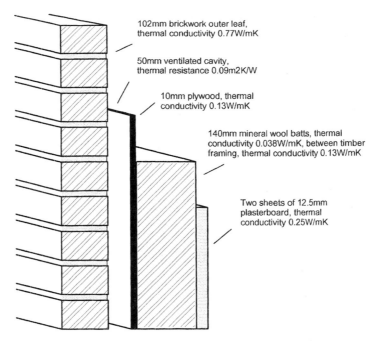

Fig. 5.5   Timber frame wall.

Table 5.8   Thermal data for timber frame wall.

Material	Thickness mm	Thermal conductivity W/mK	Thermal resistance m^{2}K/W
External surface	—	—	0.040
Outer brickwork	102	0.77	0.132
Cavity, vented	—	—	0.090
Plywood	10	0.13	0.077
Mineral wool quilt insulation	140	0.038	3.684
Timber framing	140	0.13	1.077
Plasterboard	25	0.25	0.100
Internal surface	—	—	0.130

Fig. 5.6. The resistance of each path is calculated on the basis that the materials are in series, and then the two paths are combined on the basis that they are in parallel. The first part of the calculation is illustrated in Table 5.9. The two paths are now combined in parallel to find R_{upper}:

$$\frac{1}{R_{upper}} = \frac{F_1}{R_1} + \frac{F_2}{R_2} = \frac{0.85}{4.253} + \frac{0.15}{1.646} = 0.291$$

$$R_{upper} = 3.437 \, m^2K/W$$

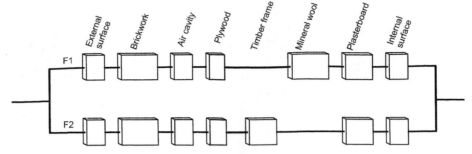

Fig. 5.6   Upper resistance limit – timber frame wall.

Table 5.9   Calculation of the upper resistance limit, timber frame wall.

	Thermal resistance, m²K/W	
	Path 1	Path 2
External surface resistance	0.040	0.040
Resistance of brickwork	0.132	0.132
Resistance of cavity	0.090	0.090
Resistance of plywood	0.077	0.077
Resistance of mineral wool quilt	3.684	—
Resistance of timber	—	1.077
Resistance of plasterboard	0.100	0.100
Internal surface resistance	0.130	0.130
Total thermal resistance of path	4.253	2.988
Fractional area of path	85% = 0.85	15% = 0.15

The lower resistance limit, R_{lower}

Each thermal bridge in the construction element is first converted to a single combined resistance. There is only one bridge in this case, formed by the mineral wool in its timber frame, as shown in Fig. 5.7. The combined resistance of the thermal bridge, R_{it} is found from:

$$\frac{1}{R_{it}} = \frac{F_{insulation}}{R_{insulation}} + \frac{F_{timber}}{R_{timber}} = \frac{0.85}{3.684} + \frac{0.15}{1.077} = 0.370$$

$$R_{it} = 2.703 \, m^2 K/W$$

This combined resistance may now be used to find the lower resistance limit, as shown in Table 5.10. Note again that R_{upper} is an overestimate of the true resistance, whereas R_{lower} is an underestimate. The average of these is very close to the true value. Hence, the total resistance of the wall is found from

$$R_T = \frac{1}{2}\left(R_{upper} + R_{lower}\right) = \frac{1}{2}(3.437 + 3.272) = 3.354 \, m^2 K/W$$

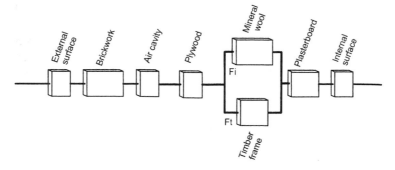

Fig. 5.7   Lower resistance limit – timber frame wall.

Table 5.10   Calculation of the lower resistance limit, timber frame wall.

	Thermal resistances, m^2K/W Thermal bridges		
	Components	**Combined**	
External surface resistance	—	—	0.040
Resistance of brickwork	—	—	0.132
Resistance of cavity	—	—	0.180
Resistance of plywood	—	—	0.077
Resistance of mineral wool (85%)	3.684 ⎱	2.703	2.703
Resistance of timber (15%)	1.077 ⎰		
Resistance of plasterboard	—	—	0.100
Internal surface resistance	—	—	0.130
Total thermal resistance, R_{lower}			3.272

and the U-value is:

$$U = \frac{1}{R_T} = \frac{1}{3.354} = 0.298 \ W/m^2K$$

Corrections to the U-value for air gaps and mechanical fixings
As in the previous example, there is the possibility of a correction for small air gaps. Again it can be assumed that the fixing of the mineral wool insulation in its timber sub-frame is such that there is no air movement on the warm side, but there are some air gaps penetrating the insulation layer. As the air gaps are in the mineral wool and timber sub-frame, $R_1 = R_{it} = 2.703$. Referring to Table 5.1, the correction for air gaps is level 1, and so $\Delta U^{11} = 0.01$. With $R_T = 3.354$, the correction is thus:

$$\Delta U_g = 0.01 \times \left(\frac{2.703}{3.354}\right)^2 = 0.006 \text{ W/m}^2\text{K}$$

As this is less than 3% of U, it may be ignored. The final U-value is rounded to two decimal places, and so the result is

$$U = 0.30 \text{ W/m}^2\text{K}$$

6 The Calculation of U-values for Ground Floors

6.1 Introduction

The accurate calculation of the U-value of ground floors is difficult and requires the rigorous procedures given in BS EN ISO 13370 [13] or in CIBSE Guide, section A3 (1999 edition) [4]. However, the full rigour of these methods may not be necessary, and AD L provides a simple approach that is adequate for most of the common constructions and ground conditions to be found in the UK. The method is based on precalculated tabulated values. There are several points to be noted:

- For solid ground floors, if the perimeter to area ratio is less than $0.12\,\mathrm{m/m^2}$, the U-value will normally be $0.25\,\mathrm{W/m^2K}$ or less without the need for insulation
- For suspended ground floors, if the perimeter to area ratio is less than $0.09\,\mathrm{m/m^2}$, the U-value will normally be $0.25\,\mathrm{W/m^2K}$ or less without the need for insulation
- For ground floors the U-value depends on the type of soil beneath the building; clay soil is the most typical in the UK and this is assumed to be the case in the following tables
- Where the soil is neither clay nor silt, the U-value must be calculated in accordance with BS EN ISO 13370 [13].

As the U-value of a ground floor depends on the ratio of the perimeter to the area, the rules for calculating this ratio must be observed. These are:

- Floor dimensions should be measured between the finished internal faces of the external elements of the building and must include any projecting bays
- For semi-detached houses, terraced houses, blocks of flats and similar structures, the floor dimensions can be either those of the individual unit or the whole building
- When considering extensions to existing buildings, the floor dimensions may be taken as those of the complete building including the extension
- Unheated spaces outside the insulated fabric (e.g. attached garages and porches) are excluded from the perimeter and area calculation, but the length of common wall between them must be included in the perimeter.

In addition to meeting U-value requirements, it is also important that the floor design should prevent excessive thermal bridging at the floor edge. This is to reduce the risk of condensation and mould growth.

6.2   Solid ground floors

For solid ground floors with all-over insulation, the U-values in Table 6.1 apply.

Table 6.1   U-values for solid ground floors.

Perimeter to area ratio m/m²	Thermal resistance of all-over insulation, m²K/W					
	0	0.5	1	1.5	2	2.5
	U-value of solid ground floor, W/m²K					
0.05	0.13	0.11	0.10	0.09	0.08	0.08
0.10	0.22	0.18	0.16	0.14	0.13	0.12
0.15	0.30	0.24	0.21	0.18	0.17	0.15
0.20	0.37	0.29	0.25	0.22	0.19	0.18
0.25	0.44	0.34	0.28	0.24	0.22	0.19
0.30	0.49	0.38	0.31	0.27	0.23	0.21
0.35	0.55	0.41	0.34	0.29	0.25	0.22
0.40	0.60	0.44	0.36	0.30	0.26	0.23
0.45	0.65	0.47	0.38	0.32	0.27	0.23
0.50	0.70	0.50	0.40	0.33	0.28	0.24
0.55	0.74	0.52	0.41	0.34	0.28	0.25
0.60	0.78	0.55	0.43	0.35	0.29	0.25
0.65	0.82	0.57	0.44	0.35	0.30	0.26
0.70	0.86	0.59	0.45	0.36	0.30	0.26
0.75	0.89	0.61	0.46	0.37	0.31	0.27
0.80	0.93	0.62	0.47	0.37	0.32	0.27
0.85	0.96	0.64	0.47	0.38	0.32	0.28
0.90	0.99	0.65	0.48	0.39	0.32	0.28
0.95	1.02	0.66	0.49	0.39	0.33	0.28
1.00	1.05	0.68	0.50	0.40	0.33	0.28

6.2.1   Solid ground floors with edge insulation

It is often more practical to insulate a floor by means of edge insulation. The edge insulation may be laid horizontally or vertically, as shown in Fig. 6.1. Where horizontal or vertical edge insulation is used *instead of* all-over insulation, a correction factor is subtracted from the U-value for an *uninsulated* solid ground floor. The correction factor is a combination of the perimeter to area ratio and an edge insulation factor, Ψ. Thus:

$$U = U_0 - \frac{P}{A} \Psi$$

where U_0 is the value for an uninsulated floor taken from Table 6.1 (the column for zero thermal resistance), and Ψ is obtained from Table 6.2.

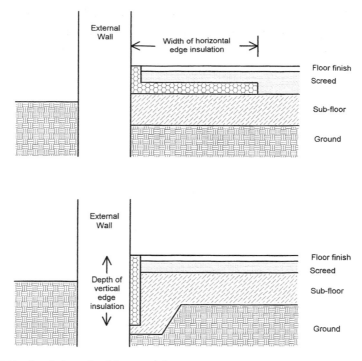

Fig. 6.1   Edge insulation of solid ground floors.

Table 6.2   Edge insulation factors for solid ground floors.

Width of horizontal insulation m	Thermal resistance of insulation, m²K/W			
	0.5	1.0	1.5	2.0
	Edge insulation factor Ψ, W/mK			
0.50	0.13	0.18	0.21	0.22
1.00	0.20	0.27	0.32	0.34
1.50	0.23	0.33	0.39	0.42
Depth of vertical insulation m				
0.25	0.13	0.18	0.21	0.22
0.50	0.20	0.27	0.32	0.34
0.75	0.23	0.33	0.39	0.42
1.00	0.26	0.37	0.43	0.48

6.2.2   Ground floors with both all-over insulation and edge insulation

There are no tables for floors with both types of insulation. For such cases, the calculation method of BS EN ISO 13370 [13] must be used.

6.3   Suspended floors

6.3.1   Uninsulated suspended ground floors

The U-values for uninsulated suspended ground floors are given in Table 6.3. These values can be used when:

- The floor deck is not more than 500 mm above the external ground level
- The wall surrounding the underfloor space is uninsulated.

Table 6.3   U-values for uninsulated suspended ground floors.

Perimeter to area ratio m/m²	Ventilation opening area per unit area of underfloor space	
	0.0015 m²/m	0.0030 m²/m
	U-value of suspended ground floor, W/m²K	
0.05	0.15	0.15
0.10	0.25	0.26
0.15	0.33	0.35
0.20	0.40	0.42
0.25	0.46	0.48
0.30	0.51	0.53
0.35	0.55	0.58
0.40	0.59	0.62
0.45	0.63	0.66
0.50	0.66	0.70
0.55	0.69	0.73
0.60	0.72	0.76
0.65	0.75	0.79
0.70	0.77	0.81
0.75	0.80	0.84
0.80	0.82	0.86
0.85	0.84	0.88
0.90	0.86	0.90
0.95	0.88	0.92
1.00	0.89	0.93

The U-values depend on the amount of ventilation which is provided to the underfloor space. This is expressed as the area in square metres of ventilation opening per unit length in metres of floor perimeter, and the table provides data for two typical values.

6.3.2 Insulated suspended floors

The U-value of an insulated suspended floor is calculated from the parameters U_o, R_f and U_f where:

U_o is the U-value of the equivalent uninsulated floor taken from Table 6.3
U_f is the U-value of the floor deck, including allowances for thermal bridging, and calculated according to the methods recommended in BS EN ISO 6946 [12], or by a numerical modelling method
R_f is the thermal resistance of the floor deck itself.

The procedure is first to obtain U_f and then to find R_f from the formula:

$$R_f = \frac{1}{U_f} - 0.17 - 0.17$$

The two values of 0.17 are the surface resistances. The U-value of the floor is then found from:

$$U = \frac{1}{[(1/U_o) - 0.2 + R_f]}$$

6.4 Example calculations

Example 6.1   Solid ground floor over clay sub-soil

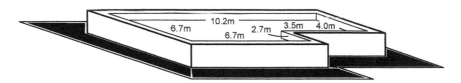

Fig. 6.2 Solid ground floor.

Figure 6.2 illustrates the floor plan of a detached house. The sub-soil is clay, and the floor is uninsulated. First, calculate the perimeter to area ratio:

P = 2 × (10.2 + 6.7) = 33.8 m
A = 10.2 × 6.7 − 3.5 × 2.7 = 58.89 m^2
P/A = 33.8/58.89 = 0.57 m/m^2

The U-value is found from Table 6.1, using the column for zero thermal resistance. The Approved Document recommends that the row nearest to the actual P/A value is used; in this case the nearest row to 0.57 is P/A = 0.55. It is not necessary to interpolate between rows because the change in U-value between rows is not large enough to warrant the extra work. The U-value of this floor is therefore:

$$U = 0.74 \, \text{W/m}^2\text{K}$$

Example 6.2   Solid ground floor over clay sub-soil with all-over insulation

The floor in Fig. 6.2 is now provided with all-over insulation between the screed and the structural floor. The insulation layer is 75 mm thick and has a thermal conductivity of 0.040 W/mK. The resistance of the insulation layer is:

$$R_{\text{ins}} = 0.075/0.040 = 1.875 \, \text{m}^2\text{K/W}$$

In Table 6.1 we again use the row for $P/A = 0.55$, but this time we must interpolate between the columns for $R_{\text{ins}} = 1.5$ and $R_{\text{ins}} = 2.0$. This interpolation is necessary because the change in U-value between columns is more significant than the change between rows. Thus:

$$\text{At } R_{\text{ins}} = 1.875, U = 0.34 - \left(\frac{1.875 - 1.5}{2.0 - 1.5}\right) \times (0.34 - 0.28) = 0.295 \, \text{W/m}^2\text{K}$$

Example 6.3   Solid ground floor over clay sub-soil with vertical edge insulation

The floor in Fig. 6.2 is provided with vertical edge insulation instead of all-over insulation. The insulation is to a depth of 750 mm, and the insulation is 75 mm thick with a thermal conductivity of 0.040 W/mK. The resistance of the insulation layer is:

$$R_{\text{ins}} = 0.075/0.040 = 1.875 \, \text{m}^2\text{K/W}$$

From Example 6.1, the perimeter to area ratio of this floor is 0.57, and its uninsulated U-value is 0.74 W/m²K. We require the edge insulation factor from Table 6.2, and it is necessary to interpolate between the columns for $R_{\text{ins}} = 1.5$ and $R_{\text{ins}} = 2.0$. Hence:

$$\Psi = 0.39 + \left(\frac{1.875 - 1.5}{2.0 - 1.5}\right) \times (0.42 - 0.39) = 0.413 \, \text{W/mK}$$

The U-value of the floor is now:

$$U = 0.74 - 0.57 \times 0.413 = 0.50 \, \text{W/m}^2\text{K}$$

Example 6.4   Uninsulated, suspended ground floor

Assume that the floor in Fig. 6.2 is an uninsulated suspended timber ground floor. The floor deck is less than 500 mm above external ground level, and the under-floor ventilation openings amount to approximately $0.0015\,\mathrm{m}^2$ per metre of floor perimeter. The perimeter to area ratio is 0.57, and taking 0.55 as the nearest value in Table 6.3, the U-value is:

$$U = 0.69\,\mathrm{W/m^2K}$$

Example 6.5   Insulated, suspended ground floor

Let the floor in Example 6.4 be insulated, with insulation fitted between the floor joists. The U-value of this floor deck may be calculated in the same way as the U-value of a wall, using the method in Chapter 5. Assuming the result of this calculation is a U-value of $0.45\,\mathrm{W/m^2K}$, the U-value of the floor is found as follows:

$$R_f = \frac{1}{U_f} - 0.17 - 0.17 = 2.22 - 0.34 = 1.88\ \mathrm{m^2K/W}$$

The uninsulated U-value, from Example 6.4, is $U_o = 0.69$, and so:

$$U = \frac{1}{[(1/U_o) - 0.2 + R_f]} = \frac{1}{[1.45 - 0.2 + 1.88]} = 0.32\ \mathrm{W/m^2K}$$

7 Compensation Calculations for Glazing

7.1 Introduction

The elemental method of both AD L1 and AD L2 allows any or all of the windows, doors or rooflights to have U-values which are greater than the standard maximum values, provided that compensating measures are taken. The compensation may be either a lower U-value in one element to offset a higher U-value in another, or it could be a reduction in the area of one or all of the elements. The guiding principle is that the total heat loss through windows, doors and rooflights should not exceed that which would occur when the building has the maximum allowable area of openings with maximum allowable U-values. However, in the case of dwellings, this flexibility cannot be used within the elemental method to increase the area of openings above the standard area provision of 25% of the total floor area. On the other hand, for buildings other than dwellings, the elemental method does not have this constraint. The examples and the procedure which follow here have wider applicability than those in the Approved Documents.

7.2 Example calculations

Example 7.1   Detached dwelling

A detached dwelling has a total floor area of $130\,\mathrm{m^2}$. There are two solid wood doors and a pair of glazed patio doors. There is also a rooflight. The standard area provision for the windows, doors and rooflights taken together is 25% of the floor area, i.e. $0.25 \times 130 = 32.5\,\mathrm{m^2}$. It is intended to use the whole of this allowance, and so the proposed areas for windows, doors and rooflights in the dwelling are:

- Windows $\qquad\qquad A_W = 23.7\,\mathrm{m^2}$
- External doors, wood (2 No.) $\qquad A_D = 3.8\,\mathrm{m^2}$
- External doors, glazed (2 No.) $\qquad A_D = 3.8\,\mathrm{m^2}$
- Rooflights $\qquad\qquad\qquad A_R = 1.2\,\mathrm{m^2}$
- Total area $\qquad A_W + A_D + A_R = 32.5\,\mathrm{m^2}$

The solid wooden doors have a U-value of $3.0\,\mathrm{W/m^2K}$, and the glazed patio doors have a U-value of $2.7\,\mathrm{W/m^2K}$. The rooflight has a U-value (including adjustment) of $1.9\,W/m^2K$. The windows have UPVC frames, and so the maximum average U-value for windows, doors and rooflights taken together is $2.0\,W/m^2K$.

Adjusting the window U-value
To achieve compliance, we must select glazing with an appropriate U-value. As the doors have U-values greater than 2.0, it is likely that the windows will need to have U-values that are less than 2.0. To find the required U-value for the windows, we first calculate the maximum and actual heat losses:

$$\text{Maximum heat loss} = 2.0 \times (0.25 A_F) = 0.5 A_F = 0.5 \times 130 = 65\,W/K$$
$$\text{Actual heat loss} = A_W U_W + A_D U_D + A_R U_R$$
$$\text{Actual heat loss} = 23.7\,U_W + 3.8 \times 3.0 + 3.8 \times 2.7 + 1.2 \times 1.9$$
$$= 23.7\,U_W + 23.94\ W/K$$

Equating these two heat losses gives the maximum U-value for the windows:

$$23.7\,U_W + 23.94 = 65$$
$$U_W = 1.73\ W/m^2K$$

Hence the maximum permissible U-value for the window is $1.73\,W/m^2K$. Referring to Table 4.2 it can be seen that this can be achieved with 16 mm low-E ($\varepsilon_n = 0.05$) argon filled double glazing ($U = 1.7\,W/m^2K$), or several different types of triple glazing.

Adjusting the window area
Alternatively, if argon filled units are unavailable, it may be possible to use 16 mm low-E ($\varepsilon_n = 0.05$) air filled double glazing, with a U-value of $1.8\,W/m^2K$. In this case the window area must be reduced, and the calculation to find the maximum permissible window area is:

$$\text{Actual heat loss} = A_W U_W + A_D U_D + A_R U_R$$
$$\text{Actual heat loss} = 1.8\,A_W + 3.8 \times 3.0 + 3.8 \times 2.7 + 1.2 \times 1.9$$
$$= 1.8\,A_W + 23.94\ W/K$$

Equating these two heat losses gives the maximum area of the windows:

$$1.8\,A_W + 23.94 = 65$$
$$A_W = 22.8\,m^2$$

Therefore, provided the window area is reduced from $23.7\,m^2$ to $22.8\,m^2$, the requirement will be satisfied with double glazed units of $U = 1.8\,W/m^2K$.

Example 7.2   Semi-detached dwelling

A semi-detached dwelling has a total floor area of $90\,m^2$. There are two insulated wood doors, each of area $1.9\,m^2$ and U-value $1.0\,W/m^2K$. There is no rooflight. For the windows it is intended to use 12 mm air gap air filled double glazing units in wooden frames, with a U-value (Table 4.2) of 2.8 W/m^{2}K. The standard area provision for the windows, doors and rooflights taken together is 25% of the floor area, i.e. $0.25 \times 90 = 22.5\,\text{m}^2$, and it is intended to install as much of this allowance as possible. The maximum window area is therefore 22.5 $- (2 \times 1.9) = 18.7\,\text{m}^2$. We first check the average U-value of the windows, doors and rooflights in the dwelling:

$$\text{Average U} - \text{value}, U_{WDR} = \frac{A_W U_W + A_D U_D + A_R U_R}{A_W + A_D + A_R}$$

$$U_{WDR} = \frac{18.7 \times 2.8 + 3.8 \times 1.0 + 0}{22.5} = \frac{56.16}{22.5} = 2.5\ \text{W/m}^2\text{K}$$

This is above the value given in Table 4.2 for wood frames, and so the window area must be reduced, and the calculation to find the maximum permissible window area is:

$$\text{Maximum heat loss} = 2.0 \times (0.25 A_F) = 0.5 A_F = 0.5 \times 90 = 45\ \text{W/K}$$
$$\text{Actual heat loss} = A_W U_W + A_D U_D + A_R U_R$$
$$\text{Actual heat loss} = 2.8\,A_W + 3.8 \times 1.0 + 0 = 2.8\,A_W + 3.8\ \text{W/K}$$

Equating these two heat losses gives the maximum area of the windows:

$$2.8\,A_W + 3.8 = 45$$
$$A_W = 14.7\,\text{m}^2$$

Therefore, provided the window area is reduced from 18.7 m^2 to 14.7 m^2, the requirement will be satisfied with double glazed units of U = 2.8 W/m^{2}K.

If instead of being reduced the window area is maintained at 18.7 m^2, it may be possible to achieve compliance by selecting a double glazed unit with a better thermal performance. The calculation to find the necessary U-value follows the same procedure as in the first part of the detached house example above. Alternatively, one can simply select a different double glazing unit and then check for compliance. If, say, a 12 mm air gap low-E ($\varepsilon_n = 0.15$) air filled double glazing, U = 2.2 W/m^{2}K, is installed in the windows, it is necessary to recalculate the average U-value:

$$\text{Average U} - \text{value}, U_{WDR} = \frac{A_W U_W + A_D U_D + A_R U_R}{A_W + A_D + A_R}$$

$$U_{WDR} = \frac{18.7 \times 2.2 + 3.8 \times 1.0 + 0}{22.5} = \frac{44.94}{22.5} = 2.0\ \text{W/m}^2\text{K}$$

This satisfies the relevant maximum average U-value requirement in Table 2.1, and therefore complies.

Example 7.3   An office building

An office building has a total external exposed wall area of $1875\,m^2$. It is proposed to fit metal framed windows to a total area of $732\,m^2$, and the windows will be fitted with 16 mm air gap low-E ($\varepsilon_n = 0.05$) argon filled double glazing, $U = 2.1\,W/m^2K$ (Table 4.2). The building is to have metal part-glazed personnel doors of U-value $3.3\,W/m^2K$ to a total area of $15.9\,m^2$. There are no rooflights. Check for compliance by the elemental method.

The requirements are:

- From Table 3.2, that the total area of openings does not exceed 40% of the exposed wall area, *and*
- From Table 3.1, that the average U-value of windows doors and rooflights does not exceed the figure for metal window frames, i.e. $U = 2.2\,W/m^2K$.

The total area of openings is $732 + 15.9 = 747.9\,m^2$. Thus:

$$\text{Openings as a percentage of exposed wall area} + \frac{747.9 \times 100}{1875} = 39.9\%$$

$$\text{Average U-value } U_{WDR} = \frac{732 \times 2.1 + 15.9 \times 3.3 + 0}{747.9} = \frac{1589.67}{747.9}$$

$$= 2.13\ W/m^2K$$

Taken together, these results satisfy the requirements of the elemental method for openings.

Example 7.4   An office building

For the office building described above, it is found that the proposed 16 mm air gap low-E argon filled double glazing units are unavailable, and that it is necessary to use 12 mm air gap low-E ($\varepsilon_n = 0.05$) air filled double glazing units instead. These have a U-value of $U = 2.5\,W/m^2K$ (Table 4.2). As both the windows and the doors have U-values greater than the maximum average of $2.2\,W/m^2K$, the average U-value must also be too high. Therefore compensatory measures are required, and in this case it is proposed to reduce the window area. The calculation may proceed as follows:

$$\text{Maximum area of openings} = \frac{40}{100} \times 1875 = 750\ m^2$$

$$\text{Maximum heat loss} = 2.2 \times 750 = 1650\ W/K$$
$$\text{Actual heat loss} = A_W U_W + A_D U_D + A_R U_R = 2.5 A_W + 15.9 \times 3.3 + 0$$
$$= 2.5 A_W + 52.47\ W/K$$

Equating these two heat losses gives the maximum area of the windows:

$$2.5 \, A_W + 52.47 = 1650$$
$$A_W = 639 \, m^2$$

Therefore the building will satisfy the requirements of the elemental method for openings provided that the area of glazing is reduced from $732 \, m^2$ to a maximum of $639 \, m^2$.

8 Target U-value Examples

The target U-value method can only be applied to a complete dwelling. The following examples illustrate the application of the procedure given in Chapter 2.

8.1 Semi-detached dwelling

Figure 8.1 is a plan view of the ground floor and first floor of a semi-detached dwelling. The dwelling lies approximately on a north-south axis. The windows have wood frames, with $11.8\,m^2$ on the southern face, $9.2\,m^2$ on the northern face, and $1.2\,m^2$ to the side. It is proposed to fit a mains gas fired boiler with a SEDBUK rating of 81%. The areas and U-values of the elements of the dwelling are given in Table 8.1.

The total area of openings is $26.00\,m^2$ and the total floor area is $94.08\,m^2$. The openings are therefore 27.6% of the floor area. Therefore the dwelling does not meet the requirements of the elemental method, but can be assessed using the target U-value method.

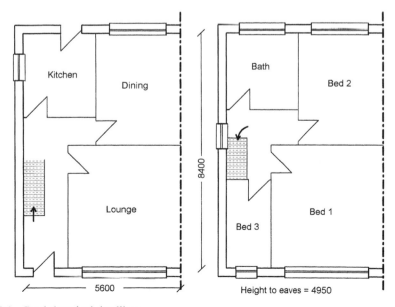

Fig. 8.1 Semi-detached dwelling.

Table 8.1 Semi-detached dwelling.

Element	Area m^2	U-value W/m^{2}K	Heat loss per degree W/K
Wall	71.00	0.35	24.85
Roof	47.04	0.20	9.41
Ground floor	47.04	0.25	11.76
Windows	22.20	1.90	42.18
Doors (2 No.)	3.80	3.00	11.40
Totals	A$_T$ = 191.08		ΣAU = 99.60

Step 1 The initial target U-value

$$U_1 = \left[0.35 - 0.19\frac{47.04}{191.08} - 0.10\frac{47.04}{191.08} + 0.413\frac{94.08}{191.08}\right] = 0.482 \text{ W/m}^2\text{K}$$

Step 2 Adjust for boiler efficiency

SEDBUK of proposed boiler = 81%
SEDBUK of reference boiler = 78% (from Table 2.2)

$$f_e = \frac{81}{78} = 1.04$$

$$U_2 = 1.04 \times 0.482 = 0.501 \text{ W/m}^2\text{K}$$

Step 3 Allowance for additional solar gain due to window frame material

The window frames are wood, and so there is no change:

$$U_3 = U_2 = 0.501 \text{ W/m}^2\text{K}$$

Step 4 Allowance for additional solar gain due to orientation of windows

South facing area   A$_S$ = 11.8 m^2
North facing area   A$_N$ = 9.2 m^2
Total window area   A$_{TG}$ = 22.2 m^2

Solar adjustment factor $\Delta S = 0.04\left[\frac{11.8 - 9.2}{22.2}\right] = 0.005$

Step 5 Convert to final target U-value

$$U_T = 0.501 + 0.005 = 0.506 \text{ W/m}^2\text{K}$$

Step 6   Find U_{AVG}

$$U_{AVG} = \frac{\Sigma AU}{A_T} = \frac{99.6}{191.08} = 0.521 \text{ W/m}^2\text{K}$$

For compliance, U_{AVG} must be less than or equal to U_T. However in this case U_{AVG} is greater than U_T, and so the dwelling does not comply by the target U-value method. There are several ways, applied either singly or in combination, in which the design of the dwelling could be altered to improve the possibility of compliance. These may either increase the target U-value, or reduce U_{AVG}, or affect both. The following are examples.

(1) *Raise boiler efficiency – and hence raise U_T*
A boiler with a higher SEDBUK rating, say 85%, could be used. The boiler efficiency factor is then $85/78 = 1.09$, and this would alter U_T as follows:

$U_1 = 0.482$
$U_2 = 1.09 \times 0.482 = 0.525$
$U_3 = U_2 = 0.525$
$U_T = 0.525 + 0.005 = 0.530 \text{ W/m}^2\text{K}$

This has resulted in U_{AVG} being less than the target U-value, and so the dwelling now complies.

(2) *Reduce the U-values of some of the elements – and hence reduce U_{AVG}*
This may be attempted by trial and error by choosing lower U-values and repeating the calculation of U_{AVG}. Alternatively, the required reduction can be estimated by setting U_{AVG} equal to U_T (which would satisfy the target U-value requirement) and calculating a new value for ΣAU.

$U_{AVG} = U_T = 0.506$
Reduced $\Sigma AU = A_T U_{AVG} = 191.08 \times 0.506 = 96.69$
Required reduction $= 99.60 - 96.69 = 2.91$

This reduction may be obtained from any one of the exposed elements, or from a combination of several. If only one element is altered, the effect on its U-value would be:

Walls	Reduction $= 2.91/71.00 = 0.04$
	New U-value $= 0.35 - 0.04 = 0.31 \text{ W/m}^2\text{K}$
Roof	Reduction $= 2.91/47.04 = 0.06$
	New U-value $= 0.20 - 0.06 = 0.14 \text{ W/m}^2\text{K}$
Ground floor	Reduction $= 2.91/47.04 = 0.06$
	New U-value $= 0.25 - 0.06 = 0.19 \text{ W/m}^2\text{K}$
Windows	Reduction $= 2.91/22.20 = 0.13$
	New U-value $= 1.90 - 0.13 = 1.77 \text{ W/m}^2\text{K}$

Doors   Reduction $= 2.91/3.80 = 0.77$
                         New U-value $= 3.00 - 0.77 = 2.23$ W/m^{2}K

Most of these reduced U-values may be difficult to achieve, with the exception of the doors. An insulated door construction providing a U-value of 2.23 W/m^{2}K or less is feasible, and would be sufficient to ensure compliance. Otherwise, if the doors cannot be altered, it may be necessary to reduce the U-value of more than one element.

(3)   *Reduce the window area – and hence reduce* U_{AVG}
From (2) above, the required reduction in ΣAU is 2.91. However, if the window area is reduced, the wall area is increased by the same amount. Consequently the gain from reducing the window area is offset by a corresponding increase in wall area. If δA is the reduction in window area, then we may write the following equation:

$2.91 = 1.90 \times \delta A - 0.35 \times \delta A = 1.55 \times \delta A$
$\delta A = 2.91/1.55 = 1.88$ m^2

Therefore:

New window area $= 22.20 - 1.88 = 20.32$ m^2
New wall area $= 71.00 + 1.88 = 72.88$ m^2

In this case, adopting a window area of 20.32 m^2 or less would be sufficient to reduce U_{AVG} to 0.506 W/m^{2}K, and hence ensure compliance.

8.2   Detached dwelling

Figures 8.2a and 8.2b are plan views of the ground and first floors of a detached dwelling. The dwelling lies approximately on a north-south axis. The windows have metal frames, with 24.8 m^2 on the southern face, 11.4 m^2 on the northern face, and 1.8 m^2 to the side. It is proposed to fit a mains gas fired boiler with a SEDBUK rating of 76%. The areas and U-values of the elements of the dwelling are given in Table 8.2.

The total area of openings is 43.70 m^2, and the total floor area is 156.00 m^2. The openings are therefore 28.0% of the floor area. The dwelling is therefore below the requirements of the elemental method, but can be assessed using the target U-value method.

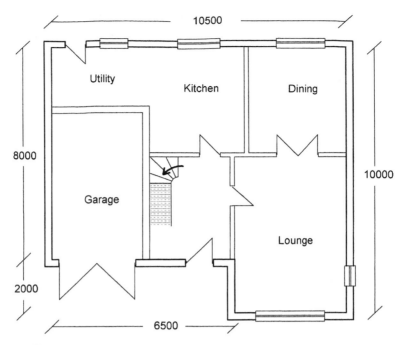

(a) Ground floor

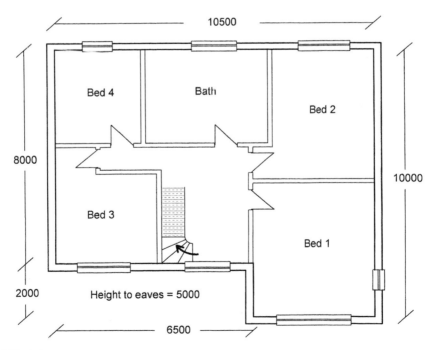

(b) First floor

Fig. 8.2 Detached dwelling.

Table 8.2 Detached dwelling.

Element	Area m^2	U-value W/m^{2}K	Heat loss per degree W/K
Wall	133.80	0.35	46.83
Roof	92.00	0.20	18.40
Ground floor	64.00	0.25	16.00
First floor to garage	28.00	0.15	4.20
Windows	38.00	2.30	87.40
Doors (3 No.)	5.70	1.70	9.69
Totals	$A_T = 361.50$		$\Sigma AU = 182.52$

Step 1 The initial target U-value

$$U_1 = \left[0.35 - 0.19 \frac{92.00}{361.50} - 0.10 \frac{64.00}{361.50} + 0.413 \frac{156.00}{361.50} \right] = 0.462 \ \text{W/m}^2\text{K}$$

Step 2 Adjust for boiler efficiency

SEDBUK of proposed boiler = 76%
SEDBUK of reference boiler = 78%

$$f_e = \frac{76}{78} = 0.974$$

$$U_2 = 0.974 \times 0.462 = 0.450 \ \text{W/m}^2\text{K}$$

Step 3 Allowance for additional solar gain due to window frame material

The window frames are metal, and so:

$$U_3 = 1.03 \ U_2 = 0.464 \ \text{W/m}^2\text{K}$$

Step 4 Allowance for additional solar gain due to orientation of windows

South facing area	A_S	= 24.8 m^2
North facing area	A_N	= 11.4 m^2
Total window area	A_{TG}	= 38.0 m^2

$$\text{Solar adjustment factor} \quad \Delta S = 0.04 \left[\frac{24.8 - 11.4}{38.0} \right] = 0.014$$

Step 5 Convert to final target U-value

$$U_T = 0.464 + 0.014 = 0.478 \ \text{W/m}^2\text{K}$$

Step 6 Find U_{AVG}

$$U_{AVG} = \frac{\Sigma AU}{A_T} = \frac{182.52}{361.50} = 0.505 \ \text{W/m}^2\text{K}$$

For compliance, U_{AVG} must be less than or equal to U_T. However in this case U_{AVG} is greater than U_T, and so the dwelling does not comply by the target U-value method. There are several ways, applied either singly or in combination, in which the design of the dwelling could be altered to improve the possibility of compliance. These may either increase the target U-value, or reduce U_{AVG}, or affect both. The following are examples.

(1) *Raise boiler efficiency – and hence raise U_T*
A boiler with a higher SEDBUK rating, say 81%, could be used. The boiler efficiency factor is then $81/78 = 1.04$, and this would alter U_T as follows:

$U_1 = 0.462$
$U_2 = 1.04 \times 0.462 = 0.480$
$U_3 = 1.03U_2 = 0.494$
$U_T = 0.494 + 0.014 = 0.508$ W/m^{2}K

This has resulted in U_{AVG} being less than the target U-value, and so the dwelling now complies.

(2) *Reduce the U-values of some of the elements – and hence reduce U_{AVG}*
This may be attempted by trial and error by choosing lower U-values and repeating the calculation of U_{AVG}. Alternatively, the required reduction can be estimated by setting U_{AVG} equal to U_T (which would satisfy the target U-value requirement) and calculating a new value for ΣAU.

$U_{AVG} = U_T = 0.476$
Reduced $\Sigma AU = A_T U_{AVG} = 361.5 \times 0.476 = 172.07$
Required reduction $= 182.52   172.07 = 10.45$

This reduction may be obtained from any one of the exposed elements, or from a combination of several. If only one element is altered, the effect on its U-value would be:

Walls   Reduction $= 10.45/133.80 = 0.08$
               New U-value $= 0.35 - 0.08 = 0.27$ W/m^{2}K
Roof   Reduction $= 10.45/92.00 = 0.11$
               New U-value $= 0.20 - 0.11 = 0.09$ W/m^{2}K
Ground floor   Reduction $= 10.45/64.00 = 0.16$
               New U-value $= 0.25 - 0.16 = 0.09$ W/m^{2}K
Windows   Reduction $= 10.45/38.00 = 0.275$
               New U-value $= 2.30 - 0.28 = 2.02$ W/m^{2}K
Doors   Reduction $= 10.45/5.7 = 1.83$
               New U-value $= 1.70 - 1.83$ negative, therefore not possible

Most of these reduced U-values may be difficult to achieve. In the case of the windows, it would be possible to achieve a U-value of 2.0 or less by specifying triple glazing. In order to retain double glazing it would be necessary to change to non-metal frames. This would reduce U_T because the factor of 1.03 for metal frames would no longer apply. Nevertheless, in this particular case, the adoption of non-metal frames may be the most practical option. Otherwise it would be necessary to take a combination of measures.

(3) *Reduce the window area – and hence reduce U_{AVG}*

From (2) above, the required reduction in ΣAU is 10.45. However, if the window area is reduced, the wall area is increased by the same amount. Consequently the gain from reducing the window area is offset by a corresponding increase in wall area. If δA is the reduction in window area, then we may write the following equation:

$$10.45 = 2.30 \times \delta A - 0.35 \times \delta A = 2.65 \times \delta A$$
$$\delta A = 10.45/2.65 = 3.94 \, \text{m}^2$$

Therefore:

New window area $= 38.00 - 3.94 = 34.06 \, \text{m}^2$
New wall area $= 133.80 + 3.94 = 137.74 \, \text{m}^2$

In this case, adopting a window area of $34.06 \, \text{m}^2$ or less would be sufficient to reduce U_{AVG} to $0.476 \, \text{W/m}^2\text{K}$, and hence ensure compliance.

9 SAP Ratings and the Carbon Index

9.1 SAP

SAP is the UK government's standard assessment procedure for rating the energy cost performance of dwellings. The SAP rating is a scale from 1 to 120. A rating of SAP = 1 is the worst possible performance (i.e. the highest energy cost) and SAP = 120 is best possible energy performance (i.e. the lowest energy cost). The scale is arbitrary, and the minimum value at which the performance of a dwelling is deemed satisfactory is a matter of opinion as expressed through government policy. The 2000 edition of AD L1 does not specify a minimum acceptable SAP, but nevertheless Building Regulations require that it should be calculated for every new dwelling, and the result made public. Details of the background to the SAP worksheets for calculating its value are given in the Government's Standard Assessment Procedure for Energy Rating of Dwellings, SAP 2001 [19].

In principle, the procedure calculates the annual energy consumed by space heating and by water heating, taking into account all relevant losses and offset by certain possible gains. This gives the energy consumption in GJ/year. The energy consumption is then converted to a cost by means of fuel price factors, to give an overall energy cost factor (ECF), in monetary units per unit floor area. This is then converted to the SAP rating according to the equation:

$$\text{SAP rating} = 97 - 100 \log_{10} (\text{ECF})$$

Clearly, the lower the energy cost, the higher the SAP rating.

9.2 Carbon factor and carbon index

One of the most important objectives of the regulation to conserve fuel and power is to control and reduce the amount of carbon dioxide released into the atmosphere by building services systems. The carbon factor and the carbon index are measures of this. The procedure is first to calculate the annual energy consumption in GJ/year. This is the same as the SAP calculation and the same worksheet is used. However, instead of being converted to a cost, the energy consumption is converted, by means of emission factors, to the total carbon dioxide, in kg/year, emitted into the atmosphere by the dwelling. The carbon emissions are then converted to carbon factor, which in turn is converted to the carbon index, by the equations:

$$\text{Carbon factor}, CF = \frac{CO_2}{(A_T + 45)}$$

$$\text{Carbon index}, CI = 17.7 - 9.0 \log_{10}(CF)$$

where CO_2 is the carbon dioxide emissions in kg/year

A_T is the total floor area of the dwelling in m^2.

It can be seen from the equations that as the CO_2 emissions fall, the carbon index rises. The carbon index is therefore used as the criterion for determining compliance under the carbon index method. The scale of values goes from 0 to 10, and the minimum value is 8.0.

9.3   Relationship between SAP and CI

The connection between SAP and CI is not consistent because one depends on fuel prices whereas the other depends on CO_2 emissions. Fuel prices are affected by market forces and taxation policy, while emissions are a fundamental property of the fuel. Nevertheless there is an approximate correspondence, and the examples in AD L1 Appendix F illustrate this. The results of these examples are given in Table 9.1, in which ε_n is the SEDBUK rating of the boiler. It may be noted that in all those examples which satisfy the requirement of the carbon index method with CI = 8 or more, the SAP rating is at least 100.

Table 9.1   Comparisons of SAP rating and carbon index.

Description	Total floor area, m^2	SAP rating	Carbon index
Two bed mid-terrace house Conventional natural gas boiler, $\varepsilon_n = 78\%$	54.6	100	8.0
Three bed semi-detached house Condensing natural gas boiler, $\varepsilon_n = 88\%$	80.0	101	8.0
Three bed semi-detached house Condensing LPG gas boiler, $\varepsilon_n = 88\%$	80.0	63	7.1
Four bed mid-storey flat Conventional natural gas boiler, $\varepsilon_n = 78\%$	90.0	107	8.5
Four bed detached house Condensing natural gas boiler, $\varepsilon_n = 90\%$	100.0	101	8.0
Two bed bungalow Condensing natural gas boiler, $\varepsilon_n = 91\%$	56.7	100	8.0

10 Example of Trade-off Calculations

The elemental method for buildings other than dwellings provides some design flexibility by means of trade-offs. There are two possible methods of trading off:

- Trade-off between construction elements by varying U-values and areas
- Trade-off between heating system efficiency and fabric performance, i.e. the areas and U-values of construction elements.

The principle that governs these trade-offs is that the actual building should be no worse than a notional building of the same size and shape which satisfies the requirements of the elemental method without employing trade-off.

10.1 Residential and conference centre

A new building is planned as a residential and conference centre. The building will be three-storey, but part of the ground floor will be two-storey in height with vehicle unloading bay doors. The building will be heated by two identical natural gas fired boilers of efficiency 78% and of combined rated heat output of 120 kW. The main characteristics of the building are:

Dimensions:	rectangular, 40 m × 15 m on plan, 10.5 m high, flat roof
Windows:	fitted with 12 mm air gap low-E ($\varepsilon_n = 0.05$) argon filled double glazing units in metal frames
	linear run of 75 m on ground and first floors
	linear run of 90 m on second floor
	total linear run 240 m
	height from sill to head 1.5 m
	total area 240 × 1.5 = 360 m^2
Personnel doors:	2 double @ 3.8 m^2, 3 single @ 1.9 m^2
	total area = 13.3 m^2
Vehicle door	1 roller door @ 25 m^2

From this we may calculate the remaining areas:

Total area of windows and personnel doors is thus 373.3 m^2
Total perimeter wall area = (40 + 40 + 15 + 15) × 10.5   = 1155.0 m^2

$$
\begin{aligned}
\text{Area of exposed wall} &= 1155 - 373.3 - 25 &&= 756.7\,\text{m}^2 \\
\text{Area of roof} &= 40 \times 15 &&= 600.0\,\text{m}^2 \\
\text{Area of ground floor} &= 40 \times 15 &&= 600.0\,\text{m}^2
\end{aligned}
$$

The area of windows and personnel doors as a percentage of the perimeter wall area is:

$$
\frac{373.3}{1155} \times 100 = 32.3\%
$$

This exceeds the allowance of 30% given in Table 3.2, and so either the window/door area must be reduced or compensating measures must be taken. This can be attempted in several ways. For example, some of the U-values for the proposed building may be lower than the maximum values specified by the elemental method, and up to half of the allowable rooflight area can be converted into an increase in window area. Consider first the U-values, as in Table 10.1.

Table 10.1   Selected U-values for proposed building.

Element	Elemental method Max U, W/m²K	Selected U W/m²K
Roof	0.25	0.25
Walls – extra insulation added	0.35	0.30
Windows – U-value determined by selected window and glazing design	2.20	2.30
Personnel door – same design as windows	2.20	2.30
Vehicle door – insulated to elemental standard	0.70	0.70
Ground floor – insulated with 75 mm EPS	0.25	0.20

In attempting to allow for the fact that both the window area and the window U-value exceed the requirements of the elemental method, the designers have decided to add insulation to the walls and the floor. The U-value of the floor may be determined from Table 6.1. The perimeter to area ratio is $110/600 = 0.183\,\text{m/m}^2$, which is nearest to 0.20 in the table, and which gives an uninsulated U-value of $0.37\,\text{W/m}^2\text{K}$. The expanded polystyrene insulation has a thermal conductivity of $0.040\,\text{W/mK}$, and so the thermal resistance of the extra insulation is $0.075/0.040 = 1.875\,\text{m}^2\text{K/W}$. Interpolating between the 1.5 and 2.0 columns of the table gives a U-value of $0.20\,\text{W/m}^2\text{K}$.

The allowable area for rooflights is 20% of the roof area, but as there are no rooflights in the proposed building, half of this allowance (i.e. 10% of the roof area) can be used to increase the allowable window area. However, this cannot be done by simply adding 10% of the roof area to the allowance for windows. This is because the maximum allowable U-values of the roof and the walls are different. This difficulty can be overcome by including rooflights in the notional building, as shown below.

The rate of heat loss from the proposed building is then compared with a notional building of the same size and shape which satisfies the elemental method. Step 1 is to prepare a heat loss table of the proposed building, incorporating the actual U-values of the proposed construction elements, as shown in Table 10.2.

Table 10.2 Proposed building heat loss table.

Element	Area m^2	U-value W/m^{2}K	Heat loss W/K
Roof	600.0	0.25	150.00
Exposed walls	756.7	0.30	227.01
Windows	360.0	2.30	828.00
Personnel doors	13.3	2.30	30.59
Vehicle loading bay doors	25.0	0.70	17.50
Ground floor	600.0	0.20	120.00
Totals	2355.0		1373.10

Average U-value, U_{AVG} = 1373.10/2355.0 = 0.583

The notional building must have the same dimensions as the proposed building, and the combined area of windows and personnel doors must account for a maximum of 30% of the perimeter wall area. However, half of the allowable rooflight area (i.e. 10% of the roof area) can be included in the notional building to help cater for the increased window area in the proposed building. The relevant areas for the notional building are thus:

Total perimeter wall area = (40 + 40 + 15 + 15) × 10.5   = 1155.0 m^2
Area of windows and personnel doors = 0.30 × 1155   = 346.5 m^2
Area of exposed wall = 1155 − 346.5 − 25   = 783.5 m^2
Area of personnel doors   = 13.3 m^2
Area of vehicle doors   = 25.0 m^2
Area of windows = 346.5 − 13.3   = 333.2 m^2
Area of rooflights = 0.10 × 600   = 60.0 m^2
Area of roof = 600 − 60   = 540.0 m^2
Area of ground floor = 40 × 15   = 600.0 m^2

Combining these with the maximum U-values allowed by the elemental method yields the heat loss table (Table 10.3). Note that the U-value of the ground floor in the notional building is 0.25 (the elemental standard) and not 0.20. This is because the value of 0.20 in the proposed building was achieved with added insulation. If it had been 0.20 *without* added insulation then it would have to be 0.20 in the notional building as well.

The results show that the proposed building has a lower heat loss rate than

Table 10.3   Notional building heat loss table.

Element	Area m²	U-value W/m²K	Heat loss W/K
Roof	540.0	0.25	135.00
Rooflights	60.0	2.20	132.00
Exposed walls	783.5	0.35	274.23
Windows	333.2	2.20	733.04
Personnel doors	13.3	2.20	29.26
Vehicle loading bay doors	25.0	0.70	17.50
Ground floor	600.0	0.25	150.00
Totals	2355.0		1471.03

Average U-value, U_{AVG} = 1471.03/2355.0 = 0.625

the notional building, and so the proposed building complies with regard to U-values and areas.

It is now necessary to consider the efficiency of the heating system. This can be done by calculating the carbon intensity of the heating system at 100% load and at 30% load, as follows:

At 100% output:

$$\text{Carbon intensity, } \varepsilon_c = \frac{1}{\Sigma R}\Sigma\left(\frac{RC_f}{\eta_t}\right) = \frac{1}{120}\left(\frac{60 \times 0.053}{0.78} + \frac{60 \times 0.053}{0.78}\right)$$

$$\varepsilon_c = 0.0679 \, \text{kg/kWh}$$

At 30% output, with only the lead boiler operating:

$$\text{Carbon intensity, } \varepsilon_c = \frac{1}{\Sigma R}\Sigma\left(\frac{RC_f}{\eta_t}\right) = \frac{1}{36}\left(\frac{36 \times 0.053}{0.78}\right)$$

$$\varepsilon_c = 0.0679 \, \text{kg/kWh}$$

Table 3.4 specifies maxima of 0.068 kg/kWh at 100% output and 0.065 kg/kWh at 30% output. The results show that the heating system is satisfactory at 100% output, but fails at 30% output. There are two ways of addressing this problem:

- Choose a more efficient lead boiler
- Check to see if boiler efficiency can be traded-off against fabric performance.

Considering the first of these options, assume that a more efficient lead boiler is available, say a condensing boiler, of efficiency 85%. The calculation at 30% load is now:

$$\text{Carbon intensity}, \varepsilon_c = \frac{1}{\Sigma R} \Sigma \left(\frac{RC_f}{\eta_t} \right) = \frac{1}{36} \left(\frac{36 \times 0.053}{0.85} \right)$$

$$\varepsilon_c = 0.0623 \, \text{kg/kWh}$$

This is below the maximum of 0.065 kg/kWh and therefore the heating system complies.

Before considering taking action to implement the second option, it should be noted that the proposed building has a lower average U-value than the notional building. This lower value may already be sufficient. Thus:

$$\text{Required average U-value}, \ U_{req} = U_{ref} \frac{\varepsilon_{ref}}{\varepsilon_{act}} = 0.625 \times \frac{0.0650}{0.0679}$$

$$U_{req} = 0.598 \, \text{W/m}^2\text{K}$$

From Table 10.2, the proposed building has an average U-value of 0.583 W/m^{2}K, and as this is less than U_{req} the heating system is satisfactory without changing either the fabric or the heating system.

11 Methods of Meeting the Lighting Standards

Paragraphs 1.41 to 1.59 of AD L2 set out the lighting efficiency standards that are required within the elemental method. The standards have two aims:

- To ensure adequate efficiency of the lamps and luminaires in converting electricity to useful light
- To provide sufficient controls to ensure that the lighting is switched on only when it is required, i.e. when there is insufficient daylight and when the space is occupied.

In any lighting design, both of these aims must be kept in mind, but for clarity of explanation, it is convenient to consider them separately.

11.1 Lamp and luminaire efficiency

The efficiency of the lighting system is measured by its *luminous efficacy*. The basic requirement is that the initial luminous efficacy of the lamp and luminaire, averaged over the whole building, should be not less than 40 luminaire-lumens per circuit-watt. This criterion must be applied to office, industrial and storage buildings, but it can also be applied to any other type of building. Note that:

- Luminaire-lumens means the useful light emitted into the space by the lamp-luminaire combination; this is less than the light emitted by the lamp because some of the light is trapped and absorbed within the luminaire
- Circuit-watt means all the electrical power consumed by the lighting circuit, and includes the power consumed by ballasts, starters, control gear, etc.
- There is no restriction on the total amount of light which may be provided; only the efficiency is controlled
- There is no restriction on the way in which the luminaires distribute light within the space
- Up to 500 W of any form of lighting is exempt, and not included in any calculations
- Display lighting, emergency lighting and specialist local lighting are considered separately from the 40 luminaire-lumens per circuit-watt requirement.

There are two methods of meeting the 40 luminaire-lumens per circuit-watt requirement:

(1) Choose lamps and luminaires which are known to have a performance which is sufficient to meet the criterion
(2) Calculate the initial luminous efficacy of the lamp and luminaire combination from photometric and electrical data.

Method 1 requires all luminaires in the building to have a light output ratio (LOR) of 0.6 or above *and* all luminaires to be fitted with lamps of any of the types given in Table 11.1. Note that in Appendix F of AD L2 this table refers to non-daylit areas of these buildings. However, as there is no separate guidance, it must be presumed that the table applies to daylit areas as well.

Table 11.1 Lamps for offices, industrial and storage buildings.

Lamp type	Lamp rating
High pressure sodium	All ratings above 70 W
Metal halide	All ratings above 70 W
Tubular fluorescent	All 26 mm diameter (T8) lamps and all 16 mm diameter (T5) lamps rated above 11 W, provided that all these lamps have low-loss or high frequency control gear
Compact fluorescent	All ratings above 26 W

Method 2 is used when either the luminaires or the lamps do not meet the requirements of method 1. In this case, it is necessary to calculate the initial luminaire efficacy, η_{lum}, using the formula given previously. This will be illustrated by example 11.1 below.

For all buildings other than offices, industrial units and storage buildings, it is occasionally impossible to meet the 40 luminaire-lumens per circuit-watt criterion because it may be necessary to use luminaires of unknown light output ratio, or to use less efficient lamps. In such circumstances, neither method 1 nor method 2 can be applied, and the alternative criterion of 50 lamp-lumens per circuit-watt must be used instead. The efficacy is 50 rather than 40 in order to compensate for the fact that, by specifying lamp-lumens instead of luminaire-lumens, the performance of the luminaires is being ignored. There are two methods of meeting the criterion:

● Calculate the average lamp-lumen efficacy from the lamp data
● Ensure that at least 95% of the installed lighting capacity uses lamps whose circuit efficacies are no worse than those given in Table 3.7.

These two methods are illustrated in Examples 11.2 and 11.3 below.

11.2   Lighting controls

The main text of AD L2 gives guidance on the selection of controls. In addition, when calculating η_{lum} it is necessary to know the control strategy in order to select a value for the luminaire control factor.

11.3   Example calculations

Example 11.1   Calculation of the average luminaire efficacy, η_{lum}

A building consists of a manufacturing unit, a storage area and offices. The offices are two storey, and the ceiling height in the other areas corresponds to the full height of the offices. The relevant data for the lighting design is:

- Manufacturing unit
  - Usage – 7-day shift patterns
  - Non-daylit – electric light only
  - Controls – timed switching according to manufacturing shift patterns
- Storage area
  - Usage – occasional use
  - Non-daylit – electric lighting only
  - Controls – manual switch on, automatic switch off by local absence detection
- Offices
  - Usage – normally daytime only
  - Daylit – window area 30% of office external wall area, glazed with clear low-E double glazing units; furthest luminaire less than 6 m from window
  - Controls – local infrared switches
- Entrance foyer, corridors and toilets
  - Usage – as for office
  - Non-daylit – electric lighting only
  - Controls – automatic on and off by occupancy sensing.

Comparing the specification for the lighting controls with the requirements of paragraphs 1.56 and 1.58 shows that they meet the requirements. Note, however, that if the offices had been glazed with tinted glass having a normal light transmittance 40%, the effective window area would be reduced to 30 × 40/70 = 17.1%. As this is less than 20%, the offices would not have been considered to be daylit.

   The efficiency of the lamps and luminaires can be checked by calculating the overall luminaire efficacy, and comparing it with the required minimum of 40 luminaire-lumens per circuit-watt. Table 11.2 illustrates a convenient format for collecting the data and carrying out the calculation. The values for luminaire control factors are derived from Table 3.6 according to the control strategy of

Table 11.2   Lamp schedules for manufacturing and office building.

	Lamp schedule				
	1	**2**	**3**	**4**	**Totals**
Position	Production area	Offices	Storage	Circulation, toilets and foyer	
Description	250 W high bay metal halide	4 lamp 12 W fluorescent with aluminium Cat 2 louvres and high frequency control gear	58 W fluorescent with aluminium louvres and mains frequency control gear	24 W compact fluorescent mains frequency downlights	
Number, N	16	16	10	24	—
Circuit power per lamp, W	271	$18 \times 4 = 72$	70	32	—
Light output Φ per lamp, lm	17000	$1150 \times 4 = 4600$	4600	1800	—
Efficacy per lamp, lm/W	63	64	66	56	—
LOR of luminaire	0.80	0.57	0.60	0.40	—
Luminaire control factor, C_L	1.00	0.80	0.80	1.00	—
Total luminaire output, lm $N.\Phi.LOR/C_L$	217600	52440	34500	17280	**321820**
Total circuit power, W	4336	1152	700	768	**6956**
Overall luminaire efficacy, lm/W	50	46	49	23	**46**

each area. The table shows that the total light output from the luminaries is 321820 lumens for a total power consumption of 6956 W. This yields an overall efficacy of $\eta_{lum} = 46$ luminaire-lumens per circuit-watt. As this exceeds the minimum value of 40, the lighting design satisfies the requirements of AD L2. In addition and independently of the calculation, the building can have a further 500 W of installed lighting.

Although it was not necessary in this case, it is useful to calculate the efficacy of each lamp and the overall luminaire efficacy of each luminaire and lamp combination. These results are included in Table 11.2. It can be seen that all the selected lamps have efficacies above 50 lm/W, indicating that they are all of high efficiency. However, because of losses in the luminaires, the overall luminaire efficacy of the lighting for Schedule 4 (the circulation areas, toilets and foyer) is only 23 lm/W, well below the 40 lm/W criterion. If instead of being 46 the overall result had been less than 40, then the low result for lamp Schedule 4 in Table 11.2 indicates that this is where corrective action would have to be taken.

Example 11.2   Calculation of the average lamp efficacy

A new restaurant consists of a reception/bar area, a dining area, kitchens and circulation/toilet areas. It is desired to use decor lighting appropriate to the building's function, and so, in the bar and dining areas, it is proposed to use a combination of concealed perimeter lighting and local lighting over tables. Controls for the reception and dining areas will be by local switching from behind the bar, and lighting to all other areas will be by local switching. The building is neither an office, an industrial unit nor a storage building, and so compliance will be attempted by calculating the average lamp efficacy. Table 11.3 gives details of the lamp schedule for the whole building, and calculates the required results.

The average lamp efficacy is 45.0 lumens per circuit-watt. This is below the criterion of 50, and so the lighting scheme does not meet the requirements of AD L2. Inspection of the results for each lamp type shows that the very low efficacy

Table 11.3   Lamp schedules for restaurant.

| | Lamp schedule | | | | |
	1	2	3	4	Totals
Position	Over dining tables	Concealed perimeter and bar lighting	Toilets and circulation	Kitchens	
Description	60 W tungsten	32 W T8 fluorescent high frequency ballast	18 W compact fluorescent mains frequency ballast	50 W T8 fluorescent high frequency ballast	
Number	30	22	8	8	—
Circuit power per lamp, W	60	36	23	56	—
Light output per lamp, lm	710	3300	1200	5200	—
Total circuit power, W	1800	792	184	448	**3224**
Total lamp light output, lm	21300	72600	9600	41600	**145100**
Efficacy per lamp, lm/W	11.8	91.7	52.2	92.9	**45.0**

for the tungsten lamps over the dining tables is responsible for the poor overall result. One solution would be to replace the tungsten lamps with compact fluorescent lamps of similar light output. The nearest would be 11 W rated lamps with a circuit power of 14 W and a light output of 690 lumens. Recalculating the table with these lamps over the dining tables raises the overall average lamp efficacy from 45.0 to 78.4 lumens per circuit-watt, thus complying with AD L2.

Example 11.3   Calculation of installed circuit power

A new sports hall consists of a sports area for badminton courts, changing room/toilet facilities, an entrance/reception area, and an office. The lighting schedule is:

- Sports playing area: 8 No. 100 W high pressure sodium downlighters
- Changing room/toilets: 12 No. 15 W compact fluorescent lamps
- Entrance/reception area: 10 No. ceiling recessed 50 W tungsten halogen downlighters
- Office: 2 No. 85 W 38 mm diameter, 2400 mm long tubular fluorescent.

Controls for the playing area, entrance/reception area and the office will be by manual control from the office. Controls for the changing room/toilets will be by occupancy detector. Table 11.4 gives details of the lamp schedules and calculations.

Only the entrance/reception area is fitted with lamps which are not listed in Table 3.7. The percentage of circuit-watts consumed by these lamps is (500/1902) × 100 = 26.3%, and so only 73.7% of the installed circuit-watts is from the list of approved light sources. It is therefore necessary to choose different lamps for the entrance/reception area.

Table 11.4   Lamp schedules for sports hall.

| | Lamp schedule | | | | |
	1	2	3	4	Totals
Position	Sports playing area	Changing rooms and toilets	Entrance and reception areas	Office	
Description	100 W high pressure sodium	18 W compact fluorescent	50 W tungsten halogen	85 W T12 linear fluorescent	
Number	8	12	10	2	—
Circuit power per lamp, W	115	23	50	103	—
Total circuit power, W	920	276	500	206	**1902**

Note that the concession allowing an additional uncontrolled 500 circuit-watts of installed lighting applies only when using the 40 luminaire-lumens per circuit-watt criterion. It cannot be applied to the 500 W of tungsten halogen lamps proposed for the entrance and reception areas.

12 CPR Calculations – Methods for Office Buildings

The carbon performance rating (CPR) calculation method is a simple technique, derived from three sources, for assessing the amount of carbon emitted into the atmosphere in units of $kgC/m^2/year$. It is used within the elemental method for assessing the contribution to carbon emissions arising from the operation of mechanical ventilation and air conditioning systems. It is also used in the whole-building method for offices in order to assess the total emissions due to mechanical ventilation, air conditioning, heating and lighting.

12.1 Origins of the CPR method for office buildings

The three principal sources relevant to the CPR method are:

- Energy Consumption Guide 19 *Energy use in offices* (ECON19) [54]. This provides benchmarks for energy consumption which have been derived from surveys of operational office buildings.
- *Energy assessment and reporting methodology (EARM): office assessment method* [45]. This provides a technique for estimating operational energy consumption, and comparing actual performance with ECON 19 benchmarks.
- CIBSE Guide *Energy efficiency in buildings* [36]. This provides a means of comparing services design with benchmarks of installed load and energy use.

The EARM has now been extended to include:

- Banks and agencies assessment method
- Hotels assessment method
- Mixed-use buildings assessment method.

12.2 The carbon performance rating for mechanical ventilation, $CPR_{(MV)}$

The value of $CPR_{(MV)}$ is found from four components:

- PD   The total installed capacity of the fans which provide mechanical ventilation. This is expressed as the sum of the input kW ratings per square metre of the floor area of the treated space, kW/m^2.

- HD   The typical annual equivalent hours of full load operation, assumed to be 3700 hours per year.
- CD   The carbon emission factor, in kgC/kWh, for the fuel used to power the fans. This is nearly always electricity, for which CD = 0.113.
- FD   A plant operating efficiency factor, which depends on provisions made to improve annual efficiency or on measures which reduce annual hours of use.

The equation for $CPR_{(MV)}$ is:

$$CPR_{(MV)} = PD \times HD \times CD \times FD$$

With HD = 3700 hours, and CD = 0.113, the equation for electrically powered fans is:

$$CPR_{(MV)} = 418 \times PD \times FD$$

12.3   The carbon performance rating for air conditioning and mechanical ventilation, $CPR_{(ACMV)}$

The value of $CPR_{(ACMV)}$ is found from PD, HD, CD and FD, as defined above, plus four more components:

- PR   The total installed capacity of the plant that provides cooling or refrigeration. This is expressed as the sum of the input kW ratings per square metre of the floor area of the treated space, kW/m^2.
- HR   The typical annual equivalent hours of full load operation of the cooling or refrigeration plant, assumed to be 1000 hours per year.
- CR   The carbon emission factor, in kgC/kWh, for the fuel used to power the cooling or refrigeration plant. This is most often electricity (CD = 0.113) or natural gas (CD = 0.053).
- FR   A plant control and management factor, which depends on provisions made to improve annual efficiency or on measures which reduce annual hours of use.

The equation for $CPR_{(ACMV)}$ is:

$$CPR_{(ACMV)} = (PD \times HD \times CD \times FD) + (PR \times HR \times CR \times FR)$$

12.3.1   Values for PD, PR, CD and CR

Values for PD and PR are determined by the ratings of the installed equipment. Values for CD and CR are taken from the carbon emission factors given in Table 3.5.

12.3.2   Values for FD

Values for FD depend on a combination of plant management features and monitoring and reporting features, and must be obtained from Table 12.1. In this table, values are selected from the most relevant column. If it is appropriate to select more than one value from that column, then the final value of FD is the product of the selected values.

Table 12.1   The factor FD.

Plant management features	Monitoring and reporting features		
	Provision of energy metering of plant and/or metering of plant hours run, and/or monitoring of internal temperature in zones, plus the ability to draw attention to out-of-range values	Provision of energy metering of plant and/or metering of plant hours run, and/or monitoring of internal temperature in zones	No monitoring provided
Operation in mixed mode with natural ventilation	0.85	0.90	0.95
Controls which restrict the hours of operation of distribution system	0.90	0.93	0.95
Efficient means of controlling air flow rate	0.75	0.85	0.95

12.3.2.1   Plant management features for FD (Table 12.1)

The plant management features in Table 12.1 should be interpreted as follows.

Operation in mixed mode with natural ventilation
If there are sufficient openable windows to provide the required internal environment by means of natural ventilation when external conditions permit, then this will only qualify as mixed mode operation if the perimeter zone exceeds 80% of the treated floor area. In addition, systems with cooling or refrigeration must have interlock controls to inhibit the air conditioning supply in zones where windows are open.

Controls which restrict the hours of operation of distribution system
This applies to controls capable of limiting plant operation to occupancy hours, with operation outside occupancy hours allowed only as necessary for efficient use of the system in order to:

- Control condensation, *or*
- Allow optimum start/stop control, *or*
- Allow the adoption of a night cooling strategy.

Efficient means of controlling air flow rate
This applies when a reduction in the air flow rate can be achieved with an efficient reduction in the input power to the fans. The reduced power may be achieved by means of motors with a variable speed control or by variable pitch fan blades. It does *not* apply to damper, throttle or inlet guide vane controls.

12.3.3   Values for FR

Values for FR also depend on a combination of plant management features and monitoring and reporting features, and must be obtained from Table 12.2.

Table 12.2   The factor FR.

Plant management features	Monitoring and reporting features		
	Provision of energy metering of plant and/or metering of plant hours run, and/or monitoring of internal temperature in zones, plus the ability to draw attention to out-of-range values	Provision of energy metering of plant and/ or metering of plant hours run, and/or monitoring of internal temperature in zones	No monitoring provided
Free cooling from cooling tower	0.90	0.93	0.95
Variation of fresh air using economy cycle or mixed mode operation	0.85	0.90	0.95
Controls to restrict hours of operation	0.85	0.90	0.95
Controls to prevent simultaneous heating and cooling in the same zone	0.90	0.93	0.95
Efficient control of plant capacity, including modular plant	0.90	0.93	0.95
Partial ice thermal storage	1.80	1.86	1.90
Full ice thermal storage	0.90	0.93	0.95

Values are selected from the most relevant column, and if it is appropriate to select more than one value from that column, then the final value of FR is the product of the selected values.

12.3.3.1   Plant management features for FR (Table 12.2)

Free cooling from cooling tower
This applies to systems that, when conditions allow, permit cooling to be obtained without the operation of the refrigeration equipment. Examples are the 'strainer cycle' and the 'thermosyphon'.

Variation of fresh air using economy cycle or mixed mode operation
This refers to systems that incorporate an economy cycle in which the mix of fresh and recirculated air is controlled by dampers, or to mixed mode operation as defined for Table 12.1.

Controls to restrict hours of operation
This applies to controls capable of limiting plant operation to occupancy hours, with operation outside occupancy hours allowed only as necessary for efficient use of the system in order to:

- Control condensation, *or*
- Allow optimum start/stop control, *or*
- Allow the adoption of a strategy to pre-cool the building overnight using outside air.

Controls to prevent simultaneous heating and cooling in the same zone
These are controls that include an interlock or dead band capable of precluding simultaneous heating and cooling in the same zone.

Efficient control of plant capacity, including modular plant
This refers to refrigerant plant capacity which is controlled on-line in such a way as to reduce input power in proportion to cooling demand while maintaining good part load efficiencies. Examples include modular plant with sequence controls, and variable speed compressors. It does *not* include hot gas bypass control.

Partial ice thermal storage
The chiller is intended to operate continuously, charging the ice store overnight and supplementing its output during occupancy.

Full ice thermal storage
The chiller only operates to recharge the thermal store overnight and outside occupancy hours.

Selecting and substituting the relevant factors in the equation gives the carbon performance rating (CPR) in $kgC/m^2/year$. This result is then compared with the maximum allowable CPR in Table 3.9. The process is demonstrated in Examples 12.1, 12.2 and 12.3 below.

12.4   The carbon performance rating and the whole-building method, CPR$_{(HLAC)}$

The whole-building CPR takes account of heating, lighting and air conditioning. In principle it is calculated in the same way as the CPR for air conditioning and mechanical ventilation, but with three additional sets of data, two for the heating system and one for the lighting. The equation for CPR$_{(HLAC)}$ is:

$$CPR_{(HLAC)} = (PD \times HD \times CD \times FD) + (PR \times HR \times CR \times FR)$$
$$+ (PB \times HB \times CB \times FB) + (PH \times HH \times CH \times FH)$$
$$+ (PL \times HL \times CL \times FL)$$

As before, the terms PD, HD, CD and FD refer to the mechanical ventilation system, and the terms PR, HR, CR and FR refer to the refrigeration plant. The additional terms are defined as follows:

- PB   The total installed capacity of the heat raising plant, kW/m^2
- HB   The typical annual equivalent hours of full load operation
- CB   The carbon emission factor, in kgC/kWh, for the fuel used in the heat raising plant
- FB   A plant management factor for the heat raising equipment.

- PH   The total installed capacity of the heat distribution system, kW/m^2
- HH   The typical annual equivalent hours of full load operation
- CH   The carbon emission factor, in kgC/kWh, for the fuel used by the heat distribution system
- FH   A plant management factor for heat distribution equipment

- PL   The total installed power of the lighting system, kW/m^2
- HL   The typical annual equivalent hours of full load operation
- CL   The carbon emission factor, in kgC/kWh, for the fuel used by the lighting system. This is nearly always electricity for which CL = 0.113
- FL   A plant management factor for the lighting system.

Values for PD, HD, CD, FD, PR, HR, CR and FR are as before. Values of PB, PH and PL are obtained from the ratings of the installed equipment, and the carbon emission factors CB, CH and CL are obtained from Table 3.5. For HB, HH and HL, typical annual equivalent hours of full load operation are:

HB = 1250 hours per annum
HH = 2500 hours per annum
HL = 2000 hours per annum

Values for the remaining parameters, FB, FH and FL are given in Tables 12.3, 12.4 and 12.5 respectively.

Table 12.3   The factor FB.

Plant management features	Monitoring and reporting features		
	Provision of energy metering of plant and/or metering of plant hours run, and/or monitoring of internal temperature in zones, plus the ability to draw attention to out-of-range values	Provision of energy metering of plant and/ or metering of plant hours run, and/or monitoring of internal temperature in zones	No monitoring provided
Controls which restrict the hours of operation of boiler plant	0.75	0.80	0.85
Efficient capacity control of boiler plant	0.75	0.80	0.85
Controls to prevent simultaneous heating and cooling in the same zone	0.90	0.93	0.95

Adapted with permission from BRE Digest 457 [44]

The three time and occupancy factors, TOF, are obtained from the following formulae:

$$TOF(A) = 1 - 0.15(TC + 2 \times OC - TC \times OC)$$
$$TOF(B) = 1 - 0.10(TC + 2 \times OC - TC \times OC)$$
$$TOF(C) = 1 - 0.05(TC + 2 \times OC - TC \times OC)$$

where   TC is the fraction of building fitted with time-based control
OC is the fraction of building fitted with occupancy control, i.e. either with sufficient zonal switching to allow occupants to control lighting in their own workspace, or with occupancy-sensing control.

The daylight management factor, DMF, is obtained from:

$$DMF = 1.3 - 0.3 \times DA \times (DC + 1)$$

Table 12.4   The factor FH.

Plant management features	Monitoring and reporting features		
	Provision of energy metering of plant and/or metering of plant hours run, and/or monitoring of internal temperature in zones, plus the ability to draw attention to out-of-range values	Provision of energy metering of plant and/ or metering of plant hours run, and/or monitoring of internal temperature in zones	No monitoring provided
Controls which restrict the hours of operation of heating system	0.85	0.90	0.95
Efficient capacity control of distribution system	0.90	0.93	0.95

Adapted with permission from BRE Digest 457 [44]

Table 12.5   The factor FL.

Plant management features	Monitoring and reporting features		
	Provision of energy metering of lighting systemand/or metering of hours of use, plus the ability to draw attention to out-of-range values	Provision of energy metering of lighting system and/or metering of hours of use	No monitoring provided
Time-based lighting control and occupancy based lighting control	TOF(A)	TOF(B)	TOF(C)
Daylight-based lighting control and ability to utilise daylight	Daylight management factor, DMF		

Adapted with permission from BRE Digest 457 [44]

where   DC is the fraction of building fitted with daylight control
DA is the fraction of building which falls within the definition of a daylit space, i.e. any space within 6 m of a window wall provided that the glazed area is at least 20% of the internal area of the window wall, or a roof-lit space with a glazing area at least 10% of the floor area.

As with FD and FR, the final value of FB, FH or FL is found by multiplying together all applicable factors in the relevant column.

12.5   Example calculations

Example 12.1   Mechanical ventilation

In a new mechanically ventilated office building the total floor area treated by the ventilation system is $4200 \, m^2$. The input power rating of the fans will be 80 kW. The fans will be driven by variable speed motors, with the speed being controlled from CO_2 sensors in the exhaust ducts. The controller will include a time function to limit operation to occupancy hours, but there is no separate metering of the ventilation plant. The required data is therefore:

PD = 80/4200 = 0.019 kW/m^2
HD = 3700 hours
CD = 0.113 kgC/kWh for electricity
FD   As there is no monitoring, FD must be chosen from the third column of Table 12.1. Two of the parameters in this column are relevant: the parameter for controls which restrict the hours of operation, and the parameter for efficient means of controlling air flow. Hence:

FD = 0.95 × 0.95 = 0.9025

The CPR for the mechanical ventilation system in this building is thus:

$$CPR_{(MV)} = PD \times HD \times CD \times FD = 0.019 \times 3700 \times 0.113 \times 0.9025$$
$$CPR_{(MV)} = 7.17 \, kgC/m^2/year$$

This must be compared with the maximum allowable CPR from Table 3.9. As the maximum for a new building is 6.5, this mechanical ventilation system fails the requirement. The result could be improved by providing separate metering of the power supplied to the fans and their control gear, and monitoring the run of plant hours. This means that the two parameters for FD would be chosen from column 2 of Table 12.1, giving:

FD = 0.93 × 0.85 = 0.7905

The revised CPR is then:

$$CPR_{(MV)} = 0.019 \times 3700 \times 0.113 \times 0.7905 = 6.28 \, kgC/m^2/year$$

This is now less than the target of 6.5, and so the mechanical ventilation system complies with the requirements of AD L2.

Example 12.2   Air conditioning

An air conditioning system is to be installed in a new air conditioned office building. The building does not have openable windows. Cooling will be provided by a pair of electrically powered speed-controlled compressors. Metering will be provided to measure the energy consumption of the refrigerant compressors, and the energy consumption of the fans used to distribute the cooled air will also be metered. Timing controls will also be provided to restrict the operation of the refrigeration and air distribution system to occupancy hours. The system data is:

Total treated floor area (TFA) $= 3600 \, \text{m}^2$
Total rated input power of compressors $= 160 \, \text{kW}$
Total rated input power of air distribution fans $= 40 \, \text{kW}$

The required data for the air distribution system is:

PD $= 40/3600 = 0.011 \, \text{kW/m}^2$
HD $= 3700$ hours
CD $= 0.113 \, \text{kgC/kWh}$ for electricity
FD   Both the compressors and the fans will be metered and so the components of FD are chosen from column 2 of Table 12.1 However, only the factor for restriction of the hours of operation applies, and so:

FD $= 0.93$

The required data for the refrigeration system is:

PR $= 160/3600 = 0.044 \, \text{kW/m}^2$
HR $= 1000$ hours
CR $= 0.113 \, \text{kgC/kWh}$ for electricity
FR   Both the compressors and the fans will be metered and so the components of FR are chosen from column 2 of Table 12.2. Two of the parameters in this column are relevant: the parameter for controls which restrict the hours of operation, and the parameter for efficient control of plant capacity. Hence:

FD $= 0.90 \times 0.93 = 0.837$

The CPR for the air conditioning system in this building is thus:

$\text{CPR}_{(ACMV)} = (\text{PD} \times \text{HD} \times \text{CD} \times \text{FD}) + (\text{PR} \times \text{HR} \times \text{CR} \times \text{FR})$
$\text{CPR}_{(ACMV)} = (0.011 \times 3700 \times 0.113 \times 0.93) + (0.044 \times 1000 \times 0.113 \times 0.837)$

$\text{CPR}_{(ACMV)} = 4.28 + 4.16 = 8.44 \, \text{kgC/m}^2/\text{year}$

This is less than the target value of $10.3 \, \text{kgC/m}^2/\text{year}$ for a new air conditioned building, and so the proposed design complies with the requirements of AD L2.

Example 12.3   Air conditioning

One area of a new office building is to be used as an archive area for the storage, retrieval and examination of ancient documents, many of which are of a delicate nature. The building will have a centralised air conditioning system, but the archive area will be treated as a separately controlled zone. Cooling will be provided by three electrically powered speed-controlled compressors. Metering will be provided to measure the energy consumption of the refrigerant compressors, and the energy consumption of the fans used to distribute the cooled air will also be metered. Timing controls will also be provided to restrict the operation of the refrigeration and air distribution system to occupancy hours in the office areas. The archive area requires continuous operation to maintain constant environmental conditions. The system data is:

Total treated floor area (TFA) = $2800 \, \text{m}^2$
Total rated input power of compressors = $180 \, \text{kW}$
Total rated input power of air distribution fans = $36 \, \text{kW}$

The archive area can be considered as a significant process load, and so its floor area and associated plant capacity can be excluded from the calculation. The data for the archive area is:

Treated floor area (TFA) = $250 \, \text{m}^2$
Estimated required input power of compressors = $22 \, \text{kW}$
Estimated required input power of distribution fans = $6 \, \text{kW}$

Subtracting the archive area data from that for the whole building gives the relevant data for the CPR calculation:

Relevant treated floor area (TFA) = $2800 - 250 = 2550 \, \text{m}^2$
Relevant rated input power of compressors = $180 - 22 = 158 \, \text{kW}$
Relevant rated input power of air distribution fans = $36 - 6 = 30 \, \text{kW}$

The data for the air distribution system is therefore:

PD = $30/2550 = 0.012 \, \text{kW/m}^2$
HD = 3700 hours
CD = 0.113 kgC/kWh for electricity
FD   Both the compressors and the fans will be metered and so the components of FD are chosen from column 2 of Table 12.1. Two of the parameters in this column are relevant: the parameter for controls which restrict the hours of operation, and the parameter for efficient means of controlling air flow. Hence:
FD = $0.93 \times 0.85 = 0.7905$

The required data for the refrigeration system is:

PR $= 158/2550 = 0.062$ kW/m^2
HR $= 1000$ hours
CR $= 0.113$ kgC/kWh for electricity
FR Both the compressors and the fans will be metered and so the components of FR are chosen from column 2 of Table 12.2. Two of the parameters in this column are relevant: the parameter for controls which restrict the hours of operation, and the parameter for efficient control of plant capacity. Hence:

FD $= 0.90 \times 0.93 = 0.837$

The CPR for the air conditioning system in this building is thus:

$$CPR_{(ACMV)} = (PD \times HD \times CD \times FD) + (PR \times HR \times CR \times FR)$$
$$CPR_{(ACMV)} = (0.012 \times 3700 \times 0.113 \times 0.7905) + (0.062 \times 1000 \times 0.113 \times 0.837)$$
$$CPR_{(ACMV)} = 3.97 + 5.86 = 9.83 \text{ kgC/m}^2/\text{year}$$

This is less than the target value of 10.3 kgC/m^2/year for a new air conditioned building, and so the proposed design complies with the requirements of AD L2.

Example 12.4   Whole-building CPR for a new air conditioned building

A new 10 storey office building is a conventional rectangular shape, 45 m × 15 m on plan, and 35 m in height. The office areas are arranged on either side of corridors which run centrally along the major axis of the building, so that the inner wall of all offices is within 6 m of the window wall. The total gross floor area is 6750 m^2, and the treated occupied area is 90% of this total. Glazing is restricted to the two main facades and is 40% of the area of these facades. The U-values are 0.3 W/m^{2}K for the external walls, an average of 2.0 W/m^{2}K for the glazing, 0.25 W/m^{2}K for the roof, and 0.2 W/m^{2}K for the ground floor.

Heating and cooling is provided via ceiling mounted cassettes, with hot water and chilled water supplied from a central plant room. Ventilation air is provided separately via a ducted system. The ventilation fans are fitted with variable speed drives which are controlled between a low (night-time) setting and maximum according to CO$_2$ sensors in the exhaust ducts. Heating and hot water is provided by three identical gas fired boilers, and cooling is provided by a pair of chiller sets.

The lighting system is conventional, providing 500 lux in all areas, with daylight control in the office spaces (approximately 80% of the occupied area), and with time control in other areas. The total installed power of the lighting system is 110 kW. All equipment, space temperatures, etc. are controlled by a BMS system, which provides full monitoring including the ability to flag out-of-range values. However, the circulating pumps for hot water are not separately monitored from their parent equipment. There is no specific control to restrict hours of operation, as out-of-hours working is expected in parts of the building. The relevant details are:

Heating and hot water:	3 identical gas-fired boilers, each rated at 60 kW
Hot water distribution pumps:	total rated input power 20 kW
Lighting:	total installed load 110 kW
Ventilation:	total installed fan power 40 kW
Refrigeration:	2 identical chiller sets, each rated at 150 kW, including heat rejection fans, with an additional 20 kW for pumps supplying chilled water to the cassettes.

The CPR calculation proceeds as follows:

The treated floor area is 90% of 6750, i.e. $6075 \, m^2$.

The boiler system

Total boiler input power = $3 \times 60 = 180 \, kW$

$PB = 180/6075 = 0.0296 \, kW/m^2$

$HB = 1250$ hours

$CB = 0.053 \, kgC/kWh$ for gas

As full monitoring is installed, values are chosen from the first column. As there are three sequenced boilers there is efficient control of boiler capacity, and the cassettes are wired to prevent simultaneous heating and cooling. The two relevant factors are 0.75 and 0.90. Therefore:

$FB = 0.75 \times 0.90 = 0.675$

The heat distribution system

$PH = 20/6075 = 0.0033 \, kW/m^2$

$HH = 3700$ hours

$CH = 0.113 \, kgC/kWh$ for electricity

The pumps are not directly monitored, and so

$FH = 0.95$

The lighting system

$PL = 110/6075 = 0.0181 \, kW/m^2$

$HL = 3000$ hours

$CL = 0.113 \, kgC/kWh$ for electricity

As there is full monitoring via the BMS, the equation for time and occupancy based control is TOF(A). The office areas, which account for 80% of floor area, are all occupancy controlled, and so $OC = 0.80$. The remaining 20% of floor area is timed, and so $TC = 0.20$. Thus:

$TOF(A) = 1 - 0.15(0.2 + 2 \times 0.8 - 0.2 \times 0.8) = 0.754$

The 80% of the floor area which is office space is all within 6 m of a window wall that is glazed to 40% of its area. These areas are therefore within the definition of being daylit, and they are also fitted with daylight control. Hence $DA = 0.8$ and $DC = 0.8$, and so:

$$DMF = 1.3 - 0.3 \times 0.8(0.8 + 1) = 0.868$$
$$FL = 0.754 \times 0.868 = 0.654$$

The air distribution system

The total installed fan power is 40 kW
$$PD = 40/6075 = 0.0066 \text{ kW/m}^2$$
$$HD = 3700 \text{ hours}$$
$$CD = 0.113 \text{ kgC/kWh for electricity}$$
There is no contribution from natural ventilation and the controls do not restrict hours of operation. On the other hand, the variable speed drives provide efficient control of air flow rate, and so:
$$FD = 0.75$$

The refrigeration system
$$PR = (300 + 20)/6075 = 0.0527 \text{ kW/m}^2$$
$$HR = 1000 \text{ hours}$$
$$CR = 0.113 \text{ kgC/kWh for electricity}$$
The controls prevent simultaneous heating and cooling in the same zone, and provide efficient control of plant capacity, and so:
$$FR = 0.9 \times 0.9 = 0.81$$

The CPR calculation can be arranged most conveniently in tabular form as in Table 12.6. The result of CPR $= 13.54 \text{ kgC/m}^2$/year is well within the target of 18.5 kgC/m^2/year for a new air-conditioned office, and so the building complies. The final column of the calculation table is useful in identifying the most significant contributions to the result. In this case, the refrigeration and lighting systems were the largest. In the case of a building which failed the target, the final column would help to identify those systems which need attention in order to improve the result.

Table 12.6   Whole-building CPR calculation.

		P kW/m²	H hours	C kgC/kWh	F	P × H × C × F kgC/m²/yr
Boiler system	B	0.0296	1250	0.053	0.675	1.324
Heat distribution system	H	0.0033	3700	0.113	0.950	1.311
Lighting system	L	0.0181	3000	0.113	0.654	4.013
Cool/vent dist. system	D	0.0066	3700	0.113	0.750	2.070
Refrigeration system	R	0.0527	1000	0.113	0.810	4.824
					CPR =	13.542

13 Solar Overheating Calculations

Appendix H of AD L2 describes a calculation method suitable for the second of the three criteria for assessing solar overheating: by specifying a maximum solar heat load per unit floor area of $25\,\mathrm{W/m^2}$. In order to carry out the calculation, the building should be divided into zones and the calculation carried out for each zone. The result for every zone must be no greater than the maximum of $25\,\mathrm{W/m^2}$.

13.1 Definitions

13.1.1 Perimeter zones

A perimeter zone is the space between a window wall (or walls) and a boundary drawn up to a maximum of 6 m from the wall (or walls). All windows, part windows, rooflights and part rooflights within a perimeter zone must be included in the calculations.

13.1.2 Interior zones

An interior zone is the space between the internal boundaries of perimeter zones and non-window walls, or between the internal boundaries of two or more perimeter zones. All rooflights and part rooflights within an interior zone must be included in the calculations.

13.1.3 Parameters and equations

The parameters used are as follows:

Q_{slw} the solar load from windows, per unit of the zone floor area $(\mathrm{W/m^2})$
Q_{slr} the solar load from rooflights, per unit of the zone floor area $(\mathrm{W/m^2})$

It is necessary to define Q_{slw} and Q_{slr} separately because they require different equations for their evaluation.

A_z the floor area of a zone $(\mathrm{m^2})$
A_g the area of glazed opening in the window wall(s) of a perimeter zone $(\mathrm{m^2})$

A_r the total area of rooflight(s) in a perimeter or interior zone (m^2)

q_{sw} the solar load due to a window, selected from Table 13.1 (W/m^2 of glazing)

q_{sr} the solar load due to a rooflight or similar horizontal opening (W/m^2 of the area of the rooflight or opening)

f_c the correction factor for the glazing/blind combination, selected from Table 13.2

f_{rw} the framing ratio for windows (default value $f_{rw} = 0.1$ for vertical windows)

f_{rr} the framing ratio for rooflights (default value $f_{rr} = 0.3$ for horizontal rooflights)

The equations for the solar loads are then:

Windows
$$Q_{slw} = \frac{1}{A_z} \Sigma A_g q_{sw} f_c (1 - f_{rw})$$

Rooflights
$$Q_{slr} = \frac{1}{A_z} A_r q_{sr} f_c (1 - f_{rr})$$

13.2   Sources of data for the parameters

13.2.1   Values of q_{sw} and q_{sr}

The solar loads, q_{sw} and q_{sr}, are averages between 07.30 in the morning and 17.30 in the evening. A single value of q_{sr} is suitable for all horizontal surfaces (i.e. rooflights):

$$q_{sr} = 327 \, \text{W/m}^2$$

For vertical surfaces (i.e. windows) the value of q_{sw} depends on orientation and must be obtained from Table 13.1.

Table 13.1   Average solar load, windows.

Orientation	q_{sw} W/m^2
N	125
NE/NW	160
E/W	205
SE/SW	198
S	156

13.2.2   Values of f_c

The correction factor for the glazing/blind combination, f_c, may either be selected from Table 13.2, or obtained from the appropriate shading coefficients.
    The shading coefficients for various glazing and shading device combinations

Table 13.2 Correction factor for glazing/blind combinations.

Glazing/blind combination (from inside to outside)	Correction factor f_c
Blind/clear/clear	0.95
Blind/clear/reflecting	0.62
Blind/clear/absorbing	0.66
Blind/low-e/clear	0.92
Blind/low-e/reflecting	0.60
Blind/low-e/absorbing	0.62
Clear/blind/clear	0.69
Clear/blind/reflecting	0.47
Clear/blind/absorbing	0.50
Clear/clear/blind/clear	0.56
Clear/clear/blind/reflecting	0.37
Clear/clear/blind/absorbing	0.39
Clear/clear/blind	0.57
Clear/clear/clear/blind	0.47

are often provided by manufacturers. For the purposes of Appendix H they are defined as:

S_c The ratio of the instantaneous solar heat gain through the glazing and its shading device at normal incidence to the instantaneous solar heat gain through clear unshaded 4 mm glass.

S_{cf} As S_c, but for glazing with a fixed shading device.

S_{ctot} As S_c, but for glazing with fixed and moveable shading devices.

There are three equations which relate these shading coefficients to the correction f_c:

- For fixed shading, including units with absorbing or reflecting glass:

$$f_c = \frac{S_c}{0.7}$$

- For moveable shading:

$$f_c = \frac{1}{2}\left(1 + \frac{S_c}{0.7}\right)$$

- For a combination of fixed and moveable shading:

$$f_c = \frac{S_{cf} + S_{ctot}}{1.4}$$

These equations maybe used to find f_c, or they may be inverted in order to determine the shading coefficient, and hence the glazing/shading device combination, which is required to meet the $25\,W/m^2$ criterion.

13.2.3   Values of f_{rw} and f_{rr}

If there is no information on the framing ratio of a window, or it is not possible to evaluate it from the dimensions, the default values should be used.

13.3   Example calculation

Part of the top floor of a college building is used as a studio for student art classes. The room is 14 m x 7 m on plan, with a floor to ceiling height of 3 m. One of the long walls is an external wall facing south-west with four windows, each 2.5 m wide by 1.5 m high. The window frames have a framing ratio of 20% of the window area. Daylight along the long internal wall is provided by a line of six horizontal rooflights, each 1 m square, with a framing ratio of 25%. The centre line of the rooflights is 1 m from the internal wall, as shown in Figure 13.1. Both the windows and the rooflights are double glazed with clear glass and an internal blind. The room can be divided into two zones. A perimeter zone can be formed by the external wall, the two end walls, and a line drawn 6 m from and parallel to the external wall. The remaining space between the perimeter zone and the inner long wall is an interior zone. Exactly half of the rooflight area falls within the perimeter zone, and the other half within the interior zone. Each zone must be considered separately.

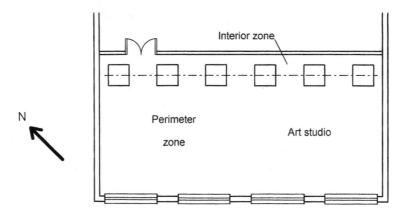

Fig. 13.1   Solar gain to a college art studio.

Perimeter zone

Area of zone	$A_z = 14 \times 6 = 84\,\text{m}^2$
Area of windows	$A_g = 4 \times 2.5 \times 1.5 = 15\,\text{m}^2$
Solar load, south-west window	$q_{sw} = 198\,\text{W/m}^2$
Glazing/blind correction factor	$f_c = 0.95$
Framing ratio for windows	$f_{rw} = 0.20$

Solar load from windows $Q_{slw} = \dfrac{1}{84}(15 \times 198 \times 0.95 \times [1 - 0.20]) = \dfrac{2257.2}{84}$

$$Q_{slw} = 26.9 \text{ W/m}^2$$

Area of rooflights $A_r = \frac{1}{2} \times 6 \times 1 \times 1 = 3\,\text{m}^2$
Solar load, rooflights $q_{sr} = 327\,\text{W/m}^2$
Glazing/blind correction factor $f_c = 0.95$
Framing ratio for rooflights $f_{rr} = 0.25$

Solar load from rooflights $Q_{slr} = \dfrac{1}{84}(3 \times 327 \times 0.95 \times [1 - 0.25]) = \dfrac{699.0}{84}$

$$Q_{slr} = 8.3 \text{ W/m}^2$$

Total solar load for perimeter zone $= 26.9 + 8.3 = 35.2\,\text{W/m}^2$

Interior zone
Area of zone $A_z = 14 \times 1 = 14\,\text{m}^2$
Area of rooflights $A_r = \frac{1}{2} \times 6 \times 1 \times 1 = 3\,\text{m}^2$
Solar load, rooflights $q_{sr} = 327\,\text{W/m}^2$
Glazing/blind correction factor $f_c = 0.95$
Framing ratio for rooflights $f_{rr} = 0.25$

Solar load from rooflights $Q_{slr} = \dfrac{1}{14}(3 \times 327 \times 0.95 \times [1 - 0.25]) = \dfrac{699.0}{14}$

$$Q_{slr} = 49.9 \text{ W/m}^2$$

Total solar load for interior zone $= 49.9\,\text{W/m}^2$

The results for both zones exceed the limiting value of $25\,\text{W/m}^2$, and so the window and rooflight design must be reconsidered. There are several possible courses of action, but in view of the room's function as an art studio, it may be inappropriate to use tinted or reflecting glass. One possible solution would be to specify, for both windows and rooflights, a double glazing system using clear glass, but with the blind fitted between the panes, for which f_c is 0.69. This, however, would not give a sufficient reduction to the result for the interior zone. Therefore, in addition, the line of rooflights could be moved an extra 250 mm away from the inner wall towards the window wall, so that three-quarters of their area is in the perimeter zone, and only one quarter in the interior zone. The revised calculations are as follows.

Perimeter zone
Area of zone $A_z = 14 \times 6 = 84\,\text{m}^2$
Area of windows $A_g = 4 \times 2.5 \times 1.5 = 15\,\text{m}^2$
Solar load, south-west window $q_{sw} = 198\,\text{W/m}^2$
Glazing/blind correction factor $f_c = 0.69$
Framing ratio for windows $f_{rw} = 0.20$
Area of rooflights $A_r = \frac{3}{4} \times 6 \times 1 \times 1 = 4.5\,\text{m}^2$

Solar load, rooflights $q_{sr} = 327\,W/m^2$
Glazing/blind correction factor $f_c = 0.69$
Framing ratio for rooflights $f_{rr} = 0.25$

Solar load from windows $Q_{slw} = \dfrac{1}{84}(15 \times 198 \times 0.69 \times [1 - 0.20]) = \dfrac{1639.4}{84}$

$$Q_{slw} = 19.5\ W/m^2$$

Solar load from rooflights $Q_{slr} = \dfrac{1}{84}(4.5 \times 327 \times 0.69 \times [1 - 0.25]) = \dfrac{761.5}{84}$

$$Q_{slr} = 9.1\ W/m^2$$
Total solar load for perimeter zone $= 19.5 + 9.1 = 28.6\,W/m^2$

Interior zone
Area of zone $A_z = 14 \times 1 = 14\,m^2$
Area of rooflights $A_r = \frac{1}{4} \times 6 \times 1 \times 1 = 1.5\,m^2$
Solar load, rooflights $q_{sr} = 327\,W/m^2$
Glazing/blind correction factor $f_c = 0.69$
Framing ratio for rooflights $f_{rr} = 0.25$

Solar load from rooflights $Q_{slr} = \dfrac{1}{14}(1.5 \times 327 \times 0.69 \times [1 - 0.25]) = \dfrac{253.8}{14}$

$$Q_{slr} = 18.1\ W/m^2$$
Total solar load for interior zone $= 18.1\,W/m^2$

The interior zone is now well within the limiting value, but the result for the perimeter zone is still too high. This suggests a variation on the above solution. If the same clear-blind-clear double glazing units are used, but the rooflights are left in their original position with their centre line 1 m from the inner wall, it is possible to manipulate the calculation by drawing the boundary of the perimeter zone 5 m from the window wall instead of the maximum permissible 6 m. The interior zone then becomes 14 m × 2 m and the rooflights fall entirely within it. The calculation becomes as follows.

Perimeter zone
Area of zone $A_z = 14 \times 5 = 70\,m^2$
Area of windows $A_g = 4 \times 2.5 \times 1.5 = 15\,m^2$
Solar load, south-west window $q_{sw} = 198\,W/m^2$
Glazing/blind correction factor $f_c = 0.69$
Framing ratio for windows $f_{rw} = 0.20$

Solar load from windows $Q_{slw} = \dfrac{1}{70}(15 \times 198 \times 0.69 \times [1 - 0.20]) = \dfrac{1639.4}{70}$

$$Q_{slw} = 23.4\ W/m^2$$
Solar load from rooflights $Q_{slr} = 0\,W/m^2$
Total solar load for perimeter zone $= 23.4 + 0 = 23.4\,W/m^2$

Interior zone

Area of zone	$A_z = 14 \times 2 = 28\,m^2$
Area of rooflights	$A_r = 6 \times 1 \times 1 = 6\,m^2$
Solar load, rooflights	$q_{sr} = 327\,W/m^2$
Glazing/blind correction factor	$f_c = 0.69$
Framing ratio for rooflights	$f_{rr} = 0.25$

Solar load from rooflights $Q_{slr} = \dfrac{1}{28}(6 \times 327 \times 0.69 \times [1 - 0.25]) = \dfrac{1015.3}{28}$

$$Q_{slr} = 36.2\,W/m^2$$
Total solar load for interior zone $= 36.2\,W/m^2$

The perimeter zone is now just within the limiting value, but the interior zone is too high. However, it can be seen that the interior zone can be brought within target by reducing the number of rooflights. The required reduction in the solar load is approximately two thirds, and so four rooflights instead of six should have the desired result. Thus for the interior zone:

Solar load from rooflights $Q_{slr} = \dfrac{1}{28}(4 \times 327 \times 0.69 \times [1 - 0.25]) = \dfrac{676.9}{28}$

$$Q_{slr} = 24.2\,W/m^2$$

Both zones are now acceptable.

14 Airtightness and Air Leakage Testing

14.1 The importance of airtightness

It has long been recognised that buildings, even when all doors, windows and other openings are closed, are not fully airtight. Wind forces create pressure differences which give rise to an unquantifiable and often unwanted air flow through a building, usually called air infiltration to distinguish it from planned ventilation. In cold weather, air infiltration leads to an additional heat load and may also cause discomfort due to cold draughts. In hot weather, it can disturb the temperature control and air distribution performance of ACMV equipment. Both of these effects make the correct sizing of heating and temperature control equipment difficult, and in the past have often led to deliberate oversizing to compensate. Furthermore, the relative importance of air leakage has increased as insulation and energy conservation standards have risen. This is because the heat loads due to air infiltration are proportionally more significant.

For example, consider a typical semi-detached house built prior to the current standards, and with an air leakage typical of much older dwellings. The heat loss due to air infiltration would be about one third or less of the total heat loss through the fabric. If the fabric insulation of the house were raised to meet the current L1 standard without changes to its airtightness, the infiltration losses and the fabric losses would be approximately equal. By sealing to the new airtightness standard required by Part L, the air infiltration loss would fall and would again be about one third of the fabric loss. In practice, it is probable that careful attention to design detail and proper supervision of on-site construction will produce dwellings with an airtightness better than the Part L requirement, thus reducing the relative importance of air leakage even further.

14.2 The mechanisms of air infiltration

Air infiltration is the combination of the driving force of air pressure differences acting against a building, and the small openings and gaps in the building fabric. A complete analysis includes the effect of internal obstructions to the infiltration flow, but for the purposes of airtightness standards, internal obstructions are ignored and only the openings and gaps in the external fabric are considered.

14.2.1   Basic equations

The openings through which ventilation air passes are too varied and complex for exact equations to be available. The usual approach is to take the normal equations for fluid flow and adapt them as necessary in an empirical fashion. There are three common types of flow.

14.2.1.1   Cracks or small openings with a typical dimension less than 1 mm

The Reynolds Number is low, and so a laminar flow type of equation may apply. For a plain pipe of radius r and length x, the flow rate Q is given by:

$$Q = \frac{\pi r^4}{8 \mu x} \cdot \Delta p^n$$

where μ is dynamic viscosity
  Δp is pressure difference between the ends of the pipe
  n is an index whose value in this equation is 1.

However, the flow paths are never as simple as a plane pipe, and the flow is rarely perfectly laminar, and so it is usually considered that a more representative version of this equation is of the form:

$$Q = C\, \Delta p^n$$

where   C is a flow coefficient, normally expressed per unit area of opening
  n is an index whose value is less than 1, and is usually about 0.67.

The flow coefficient of cracks is often expressed per unit length of crack, L, in which case

$$C = bL$$

where   b is a flow coefficient.

14.2.1.2   Openings greater than 1 mm

The Reynolds number is high and flow is turbulent. Hence an equation similar to the orifice plate formula may be used:

$$Q = C_d A_e \left[\frac{2\Delta p}{\rho} \right]^{\frac{1}{2}}$$

where C_d is a discharge coefficient
  A_e is the effective area of the opening
  ρ is the density of air.

Typical values of the constants are $C_d = 0.64$ and $\rho = 1.2 \text{ kg m}^{-3}$. Substituting these yields a simpler form of the equation:

$$Q = 0.827 \; A_e \; \Delta p^{\frac{1}{2}}$$

14.2.1.3   Ducts and chimneys

The duct flow equation is

$$Q = A \left[\frac{2A}{Efd\rho} \Delta p \right]^{\frac{1}{2}}$$

where   A is the cross-sectional area of duct
E is the cross-sectional perimeter
f is the friction coefficient
d is the duct length
ρ is the density of air

14.2.2   Driving forces

There are two independent causes of pressure difference, wind induced pressures and stack effect pressure differences. The latter are caused by internal to external temperature differences.

14.2.2.1   Wind induced pressure difference

Relative to the static pressure of the free wind, the time averaged pressure, Δp_w, acting on the surface of a building is

$$\Delta p_w = \frac{1}{2} \cdot \rho C_p v_h^2$$

where ρ is the density of air
C_p is an experimentally determined pressure coefficient
v_h is the wind velocity at a reference height, usually taken as the height of the building.

Meteorological data usually quotes wind speeds at a standard height of 10 m, and so v_h must be estimated from this meteorological wind speed. Near the earth's surface, the wind speed follows a power law:

$$v_h = v_m \; Kh^a$$

where v_m is wind velocity at a height of 10 m
v_h is wind velocity at height h

K and a are constants depending on the terrain over which the wind passes.

Typical values for K and a are given in Table 14.1.

Table 14.1 Wind speed constants.

Terrain	K	a
Open flat country	0.68	0.17
Country with scattered windbreaks	0.52	0.20
Urban	0.35	0.25
City	0.21	0.33

14.2.2.2 Stack effect pressure difference

The stack effect is the pressure due to the difference between inside and outside temperatures, over a pair of openings at different heights. The natural pressure gradient in the air due to height is different inside and outside the buildings because the air density is different. The net stack effect pressure difference, Δp_s, between the two openings is given by:

$$\Delta p_s = \rho g. 273 (h_2 - h_1) \left[\frac{1}{T_{ext} + 273} - \frac{1}{T_{int} + 273} \right]$$

where   g is the acceleration due to gravity
h_2-h_1 is the difference in height between the two openings
T_{ext} and T_{int} are the external and internal temperatures, °C.

Substituting typical values for ρ and g, and assuming that both temperatures are near the values to be expected in practice, a simpler approximate version of this formula can be obtained:

$$\Delta p_s \cong 0.043 (h_2 - h_1)(T_{int} - T_{ext})$$

In tall buildings, the stack effect can be just as significant as the effect of wind pressures. As an example, taking the difference in height between a pair of openings in a tall building to be 30 m, and the internal to external temperature difference to be 20°C, this formula yields a stack pressure difference of about 26 pascals, or about 13 pascals across each opening. However, intermediate floors will absorb much of this pressure difference and so only a small proportion of the 13 pascals will act across any one opening in the external fabric.

14.2.3 Openings and the leakage area

The resistance to the pressure differences is provided by the combination of the multitude of openings, gaps and cracks in the building fabric. Some of these will act in parallel and some in series. The effect of all the openings in a flow path can be combined into a single opening with an effective area, A_e.

Openings in parallel: The total effective opening is the sum of the separate openings:

$$A_e = A_1 + A_2 + A_3 \quad \text{or} \quad C_e = C_1 + C_2 + C_3$$

Openings in series: The effective opening is found from:

$$\frac{1}{A_e^2} = \frac{1}{A_1^2} + \frac{1}{A_2^2} + \frac{1}{A_3^2} \quad \text{or} \quad \frac{1}{C_e^{1/n}} = \frac{1}{C_1^{1/n}} + \frac{1}{C_2^{1/n}} + \frac{1}{C_3^{1/n}}$$

where   n is the index associated with C in section 14.2.1.1, whose value is about 0.67.

It is clear from the above equations that the effective area of all the openings is the critical factor in determining the air leakage rate through a building. In order to control the size of this area, it is necessary to be able to assess it. For specific parts of a construction element, say the crack around a door jamb, it is possible to obtain approximate dimensions and hence to calculate its leakage area. However, the major concern in air leakage is the multitude of non-quantifiable openings which occur unintentionally, especially at junctions between construction elements and at points where services penetrate the fabric. Consequently, the only practical way of assessing leakage area is by measurement after the building has been completed.

14.3   The measurement of air leakage

The equation for the leakage flow rate for a complete building will be similar to the equations given in section 14.2.1 above for individual components, and so may be expected to be of the form:

$$Q = k(\Delta p)^n$$

where   Q is the overall leakage flow rate
k is a constant which is a measure of the overall effective leakage area
Δp is the pressure difference across the fabric of the building
n is an index.

Examination of the formulae in 14.2 above suggests that the index n is somewhere in the range 0.5 to 1.0. Its exact value will depend on the precise combination, sizes and shapes of openings and leakage paths that exist in a particular building. Typically, for a complete building, n is usually found to be in the range 0.65 to 0.67.

In order to determine the air leakage performance of a building, it is necessary to measure both of the variables in this equation, i.e. the flow rate Q and the pressure difference Δp. There are several methods available for this, the principal one being the fan pressurisation test.

14.3.1   Fan pressurisation measurement

In the fan pressurisation method, the pressures are generated artificially by a fan which blows air into (or extracts air from) the building. The fan is connected to a convenient opening, usually a door, and the flow rate increased up to a maximum and then allowed to decrease. The fan assembly itself is either pre-calibrated, so that the flow rate is known for any given fan speed, or the fan assembly is fitted with pressure tappings or a restrictor plate from which the flow rate can be measured. The resulting internal to external pressure differences are measured with suitable pressure sensors inside and outside the building. The result is a graph of flow rate against pressure difference which, because it is a power law, is a curve. Figure 14.1 is a schematic diagram of the experimental arrangement, and shows the type of graph that is produced. Because the air flow is under direct control, and because both air flows and pressure differences are measured directly, this is a controllable and repeatable method, and as such has been chosen as the preferred method in the regulatory framework. However, the test procedure has some drawbacks:

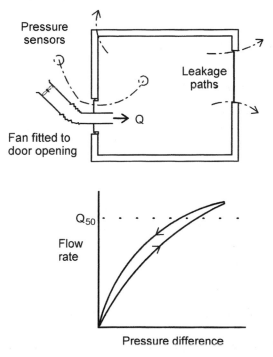

Fig. 14.1   The fan pressurisation test.

- In order to be able to make reliable measurements, the pressure differences used in the test must be sufficiently high, partly to overcome the naturally occurring pressures due to wind and stack effect, and partly to improve the accuracy of the measurements. Pressure differences up to 60 pascals are normally used, which is much higher than the maximum of about 8 to 10 pascals created by wind or stack effect.
- The curve obtained as the building is depressurised usually differs from that obtained during pressurisation, leading to some uncertainty in the interpretation of the results.
- The high pressures may cause the external fabric to leak in a way which is not normal. For example, draught excluding seals which are designed to resist an externally applied pressure may be forced away from their seating by the internally generated pressure of the fan, causing the air leakage to be exaggerated. This can be compensated by reversing the air flow and depressurising the building, so that the air flow is inwards. This is often done when testing small buildings such as dwellings, but is difficult or even impossible with large buildings.
- In a large building, the internally generated pressure distribution will vary according to location with respect to the entry point of the fan, and this variation may be significant, especially if the building contains internal walls and floors.
- In large buildings, it may be difficult to generate sufficient air flow to reach a pressure difference which can be measured with acceptable accuracy.

There are also some difficulties in interpreting the test results:

- Because of the power law nature of the relationship between air flow rate and pressure difference, the constant k in the equation is an index rather than a precise measure of the area of the openings in the external envelope.
- Measurements at the pressure differences normally generated naturally are prone to excessive error; therefore results at much higher pressure differences, around 50 pascals, are needed to give good repeatability and reliability.
- The behaviour of the leakage paths during a pressurisation test at 50 pascals may not be typical of their behaviour in normal conditions; for example, the air flow in some leakage paths may be laminar at low pressure differences, but become turbulent at the much higher pressure differences used in a test.

Etheridge and Sandberg [56] give a fuller discussion of the possible errors associated with fan pressurisation measurements.

14.3.2   Preparation for a fan pressurisation test

The pressurisation test must be carried out on the completed building. Details of the recommended test procedures and the instrumentation are given in CIBSE

TM 23 [21]. The following points relate to preparation of the building and the conditions necessary for a test:

- Determine the exact boundaries of the external envelope of the building that is to be tested. Service spaces such as boiler rooms, lift rooms or rooms housing switch gear, may be taken as outside the space to be tested. Adjoining buildings, connected for example by a corridor, may also be considered separate. In all of these cases, the space taken as being outside must be sealed from the test space.
- Switch off all combustion appliances and mechanical ventilation systems.
- Close all adjustable openings in the external fabric such as windows, doors, rooflights, ventilators, etc.
- Seal up intentional fixed openings such as air bricks, trickle ventilators, ducts to combustion equipment, chimneys and flues (though flues from room sealed appliances need not be sealed), ventilation intakes and extracts, etc.
- Ensure that all internal doors, etc. are kept open so that air within the building can move freely, in order that air pressures within the building are as uniform as possible.
- Fit and seal the fan assembly into a convenient opening, usually an external door, and install the internal/external pressure and temperature measuring devices.
- Ensure that external conditions are such that air flow due to the naturally occurring wind and stack effect forces is small enough not to upset the measurements. For this condition to be satisfied, the external wind speed should be less than $3\,\mathrm{ms}^{-1}$ and the difference between the internal and external temperatures should be less than 10°C.

When all these points have been attended to, the test itself may begin.

14.4   The air leakage criterion

14.4.1   The Part L standard

To avoid some of the difficulties of measurement and interpretation, Part L of the Building Regulations has chosen a relatively simple criterion for deciding whether or not a building is sufficiently airtight. It first assumes that the fan pressurisation method provides a more reliable test of the airtightness of a building than the tracer gas, or any other, method. It then takes the experimentally determined flow rate at a pressure difference of 50 pascals as being the most reliable measure of the air tightness of a building's external fabric. This result could then be used to calculate the constant, k, which could then be used to estimate the actual leakage area in the building envelope. However, for the purposes of regulation this is unnecessary, as the leakage flow rate is itself the quantity which matters, and so the measured flow at 50 pascals is used as the criterion of airtightness. However, the total flow rate depends not only on pressure difference but also on the surface area of the building. There are several options for making allowance for the size of a building, the most common being to divide the flow rate at 50 pascals by the external surface area of the building. There is a difference of opinion as to whether or not a solid ground floor should be included in the external surface area, as this is not expected to contribute to the leakage. There are, therefore, two indices:

The air leakage index $AI_{50} = Q_{50}/S$
The air permeability $AP_{50} = Q_{50}/S_T$

where Q_{50} is the flow rate through building envelope at an applied pressure of 50 Pa
S is the internal surface area of external facade, including all surfaces except a solid ground floor
S_T is the internal surface area of external facade, including all surfaces, *including* a solid ground floor.

Part L uses the second of these indices, the air permeability at 50 Pa, or AP_{50}, for all buildings, so that it is then possible to specify the same criterion for all buildings, whether or not the ground floor is solid. The current requirement is that all buildings should have an AP_{50} that is less than or equal to $10 m^3/h/m^2$.

14.4.2   Comparison with other standards

Several countries, especially in Scandinavia and North America, have had standards for airtightness for some time. Comparisons with standards in these and other countries are difficult because of differences in the way those standards are expressed. For example, although the 50 pascal pressure difference is commonly used, other standards may use higher or lower pressures. Also, some standards exclude solid ground floors and use the air leakage index. It has also been common to express the results in terms of effective (or equivalent) leakage area, or ELA. This is found by inverting the orifice plate formula in section 14.2.1.2, with $C_d = 1$:

$$ELA = Q \left[\frac{\rho}{2\Delta p} \right]^{\frac{1}{2}}$$

Because the relationship between Q and Δp is not linear, it is necessary to specify the pressure difference to be used in this formula, and to measure the corresponding flow rate. Typically, $\Delta p = 4$ pascals is chosen because it is representative of average weather conditions. But this pressure difference is too low to give a reliable measure of flow rate, and so Q must be found by extrapolation from measurements at higher pressures.

Nevertheless, such comparisons as can be made suggest that the new Part L standard of $10\,m^3/h/m^2$ is less stringent than the best standards elsewhere. Even within the UK, new buildings are often expected to have an AP_{50} that is well below the Part L standard. For example, CIBSE 23 [21] quotes target figures for 'best practice' which are in the range 1.5 to $5.0\,m^3/h/m^2$, according to building type. The Part L standard is therefore not excessively onerous, but even so, should make a significant contribution to improving energy conservation. This is because measurements on the current UK building stock have shown that the airtightness of UK buildings is often very poor.

14.4.3   Meeting the Part L standard

The relevant legal requirement is that reasonable provision should be made to limit heat loss (or gain) through the fabric of a building. This provision should include the limitation of air leakage. In the case of dwellings and other buildings of less than $1000\,m^2$ floor area, the Approved Documents state that compliance can be demonstrated either by adherence to approved detailing [2], or by conducting a satisfactory pressurisation test on the finished building. For buildings of greater than $1000\,m^2$ floor area, the AD L2 gives no alternative to pressurisation testing, suggesting that every large building must be tested and shown to reach the standard. However, the Approved Documents are not themselves legally binding, and it is possible that alternatives to the pressure testing of large buildings will become acceptable.

14.5   Air leakage paths

There is an expanding literature giving advice on construction details for avoiding air leakage. In addition to the references already referred to ([2], [21] and [37]), Potter [57] gives a checklist, and Perera *et al.* [58] provide details for office type buildings. The particular areas of concern can be categorised into:

- Wall components
- Roof components
- Floor components
- Windows doors and rooflights
- Joints, and materials and methods for sealing
- Penetration of the fabric by load bearing elements (steelwork, timber joists etc.)
- Penetration of the fabric by shafts (risers, lift shafts, etc.)
- Penetration of the fabric by services.

The principle behind all air leakage prevention is that the conditioned space within a building is enclosed by a continuous impermeable layer. This may be provided by the normal building components suitably sealed on surfaces and at joints, or by a continuous membrane such as polythene film. The impermeable layer will, in most cases, also be effective as a vapour barrier, and its positioning is therefore critical to avoid interstitial condensation. The general rule is that vapour barriers must be positioned on the warm side of an insulation layer.

14.5.1   Wall, roof and floor components

Many traditional materials are porous to air flow. Blockwork and brickwork are porous even when all joints are filled, and although plasterboard itself has a very low porosity, the joints between adjacent sheets will often open up after a period. Good quality rendering and/or painting can reduce air leakage to negligible levels, but these surface finishes are prone to damage and degradation. In most cases it may be necessary to include in the construction a barrier, such as a continuous membrane, specifically designed to limit air leakage. Curtain walling systems are often assembled from non-porous materials, but the fixing and jointing of the components can leave cracks and gaps which must be sealed.

Ceilings that form the underside of a roof must be airtight in the same way as the external walls. The plasterboard ceilings used in domestic construction are likely to be satisfactory, except where they are penetrated by light fittings or other services. Boarded ceilings, suspended ceiling tiles, etc. are inherently leaky to air flow, and the airtightness must be provided by some other component such as a concrete or metal roof deck, or be provided by an airtight membrane. If the adjoining wall is plastered, the plaster must be taken up to or above the ceiling airtightness barrier. Solid floors are likely to be impermeable, but suspended floors (especially timber) will almost certainly be leaky.

14.5.2   Windows, doors and rooflights

The materials are non-porous, but the sealing of edges is critical. This includes the sealing of the glass in its framing material, and seals to the reveals of openable elements. All window and door frames should be effectively sealed to the inside surface of the surrounding structure. Loading bay doors and security shutters are often porous because, for example, they are of the roller type. If their positioning is such as to be relevant and they need to be airtight, suitable designs must be selected.

14.5.3   Joints and sealing

The edges at which components meet are particularly vulnerable to leaking. In addition to the more obvious joints, such as at windows and doors, the edge junction of walls to floors and walls to ceilings must be considered. The line along which an internal wall meets an external wall can also be a weakness. The materials used for jointing can generally be classified into sealants and gaskets. Gun applied sealants of the elastic/elastomeric type are very effective provided they are applied to a well designed joint. This means that they must adhere to the surrounding materials and be able to accommodate the cyclical movements which are inherent in any joint. If the joint is badly designed, or adhesion is lost due to unsuitable materials or dirt, failure of the seal will occur very quickly. Even so, the maximum life expectancy of sealed joints is likely to be around 20 years, which is less than the design life of the building, and so easy maintenance or replacement ought to be considered in their design.

Foam sealants, usually applied from a pressurised canister, are useful for sealing large irregular gaps and also for fixing some components. They are frequently used in remedial work. While they are capable of providing a good seal, the performance and life expectancy of such joints with respect to air leakage is difficult to assess in general terms, and will probably depend on each specific case. Although foam sealants may have a role to play in sealing awkward and inaccessible holes, it would seem unwise to rely on this type of material as a primary method of achieving airtightness. Preformed flexible foams are also available, often supplied as a continuous strip. These are intended to be forced into a gap, so that the compression against the side of the gap forms a seal.

Gaskets made of neoprene or silicone have been in use for a long time and are most frequently used in curtain wall systems, and for supporting the glazing units in UPVC frames. They are also used around hatches, and supplied in kit form for sealing around piped and ducted services where these penetrate the external fabric. These gaskets are nearly always designed for a specific application and are unlikely to perform satisfactorily if used otherwise. One of the difficulties with gaskets is maintaining adequate pressure on them after they have been installed.

Draught stripping of windows and doors is now normally supplied as an integral part of the window/door assembly. The manufacturer should be able to supply evidence that his product will meet an appropriate air infiltration performance level for windows, such as BS 6375 [59].

14.5.4 Penetration of the external fabric

Penetration of the external fabric by structural members is best avoided altogether. This can be done by adopting suitable construction details, the most obvious example being to support wood joists in dwellings on joist hangers rather than directly in the external wall. Solutions may be less simple in the case of steel or reinforced concrete frame buildings, but the general principle is that the material, component or membrane which provides the necessary airtightness is continuous. The irregular profile and/or the uneven surface textures which often exist where a structural component penetrates the airtightness barrier are likely to be difficult to seal.

Some penetration by shafts and service components is usually inevitable, and the sealing arrangements must be carefully designed and installed. Recessed electrical sockets and similar components require careful attention. In large buildings lift shafts can be a particular problem, and the seals to the doors and between the shaft and the main building structure must be fully designed and specified. At the same time, a ventilation air supply for the occupants of the lift must not be forgotten.

14.6   Alternative test methods

Although the air pressurisation test is the only method offered by the Approved Documents, there may be occasions when it is difficult to apply. There a number of alternative test methods which can give a measure of airtightness, though at present none are as reliable.

The acoustic method consists of the generation of very low frequency sound waves within the building. These produce a continuously fluctuating pressure difference, whose magnitude depends on the leakage area of the building. The pressure differences are very small, and because they are alternately positive and negative, it can be argued that they represent a more realistic test then the much higher pressures of the pressurisation test. Unfortunately, the method has some drawbacks. In particular, it measures the acoustic impedance of the structure rather than its leakage area, and it is not always possible to derive the one from the other.

The tracer gas method, described in more detail in section 14.6.2 below, measures the infiltration rate of a building subjected to naturally occurring pressure differences. As the infiltration rate is really a flow rate, the test is in effect measuring the air permeability at a much lower pressure than 50 pascals used in a pressurisation. There is therefore a connection between AP_{50} and the measured infiltration rate. Unfortunately, the conversion from one to the other cannot easily be accomplished without the risk of introducing large errors.

14.6.1   Air permeability and infiltration rate

Approximate formulae relating the flow rate at 50 pascals to the measured infiltration rate at various average wind speeds have been given by Liddament [60]. These formulae apply mainly to small buildings, especially dwellings, and after some manipulation may be expressed in terms of the air permeability:

For Z measured at high average wind speeds (> 4 ms^{-1}) $\qquad AP_{50} \cong \dfrac{10VZ}{S_T}$

For Z measured at typical average wind speeds, $\qquad AP_{50} \cong \dfrac{20VZ}{S_T}$

For Z measured at low average wind speeds (< 4 ms^{-1}) $\qquad AP_{50} \cong \dfrac{30VZ}{S_T}$

where   Z is the measured infiltration rate in air changes per hour
V is the internal volume of the building.

The formula for typical average wind speeds is also given in CIBSE TM23 [21], which also quotes an adaptation of this formula for larger non-domestic buildings. In terms of air permeability, this is:

$$AP_{50} \cong \frac{V}{S}\frac{20VZ}{S_T}$$

where   S is the surface area of the walls and roof, excluding the ground floor.

Sherman and Grimsrud [61] have developed more detailed formulae relating effective (or equivalent) leakage area, ELA, to infiltration rate, with allowances for the conditions which prevail at the time of the infiltration measurement. After some manipulation, these can be adapted to give an expression for the air permeability at 50 pascals:

$$AP_{50} = \frac{Q_{50}}{S_T} = \frac{10ELA}{S_T\sqrt{\rho}} = \frac{10Z}{S_T\left(f_s^2 \times \Delta T + f_w^2 \times v^2\right)\sqrt{\rho}}$$

where f_s is a function to correct for stack effect
f_w is a function to correct for wind pressure
ΔT is the internal to external temperature difference during a test measurement
v is the measured open site wind speed at a height of 10 metres.

The formulae for f_s and f_w:

$$f_s = \left(\frac{1 + 0.5R}{3}\right) \times \left(1 - \frac{X^2}{(2 - R)^2}\right) \times \left(\frac{gH}{T_o}\right)$$

$$f_w = C(1 - R)^{0.333} \times A \times \left(\frac{H}{10}\right)^B$$

$$R = \frac{ELA_{ceiling} + ELA_{floor}}{ELA} \quad ; \quad X = \frac{ELA_{ceiling} - ELA_{floor}}{ELA}$$

where   A, B are terrain parameters
C is a shielding parameter
H is the height of the building in metres
T_o is the indoor air temperature
g is the acceleration due to gravity
$ELA_{ceiling}$ and ELA_{floor} are the equivalent leakage areas of the ceiling and ground floor planes respectively.

In many cases, measurements will be carried out on newly completed unoccupied buildings. If there is no heating or cooling in operation, ΔT will be close to zero and so the equation for AP_{50} simplifies to:

$$AP_{50} = \frac{10Z}{S_T \times f_w^2 \times v^2 \sqrt{\rho}}$$

The range of values for the coefficients used to calculate f_w are:

C	0.11 to 0.34	median 0.25
A	0.47 to 1.30	median 0.85
B	0.10 to 0.35	median 0.20

Taking, as typical, $R = 0.2$ and $H = 15$, gives the following range of values for f_w:

Minimum	$f_w = 0.05$
Median	$f_w = 0.21$
Maximum	$f_w = 0.47$

These formulae can be used to calculate the infiltration rate which is to be expected at average wind speeds for a building which just meets the $AP_{50} = 10\,m^3/h/m^2$ standard. By inverting the formulae, we have:

For dwellings $\qquad Z = \dfrac{1}{2}\dfrac{S_T}{V}$

For larger buildings $\quad Z = \dfrac{1}{2}\dfrac{S}{V}\dfrac{S_T}{V}$

For dwellings, S_T/V is about 0.8. For larger buildings, both the ratios S/V and S_T/V are likely to be 0.6 or less. Thus the infiltration rates at average wind speeds which correspond to the Part L airtightness standard are about 0.4 ach for dwellings and about 0.2 ach for larger buildings. Note that as the size of a building increases, its volume, V, tends to increase in proportion to the cube of a typical linear dimension, whereas the surface areas S and S_T increase only in proportion to the square. Hence, for a facade design with a particular air leakage characteristic, the measured infiltration, in air changes per hour, tends to reduce as the building size increases.

14.6.2   Tracer gas measurement

In a tracer gas experiment, the concentration of tracer gas released into the building is monitored over a period of time ranging from a minimum of say 30 minutes to perhaps several hours. By measuring either the decay in concentration or the rate at which tracer gas must be injected to maintain constant concentration, it is possible to determine the air flow rate, Q. Experiments of this type are normally carried out in natural conditions, so that the flow is the result of the prevailing wind and stack pressures. Thus, in order to find Δp, it is necessary to take simultaneous measurements of the wind velocity and the internal and external temperatures, and use equations such as those in sections 14.2.2.1 and 14.2.2.2 to calculate the relevant pressure differences. This is an inherently unreliable process for two reasons. Firstly, the air flow rate is not measured directly and must be inferred from the tracer gas measurement, thus introducing a degree of uncertainty. Secondly, the wind velocity and direction are rarely stable over the duration of a test, and in tall buildings where the stack effect is important, there will be temperature gradients both inside and outside the building. Even if conditions are sufficiently stable, the natural pressures may be too high or too low to produce an easily measurable air flow. On the other hand, a tracer gas measurement is performed in the normal conditions to which the building is subjected, and there may be circumstances when this type of measurement may be useful. If a pressurisation test is not possible, one or other of the formulae listed above could be used to convert the infiltration rate derived from a tracer gas measurement to an estimate of AP_{50}, but for the purposes of compliance the procedure would have to be agreed with the relevant building control body.

15 Thermal Bridges

15.1 The importance of thermal bridges

Thermal bridges are areas of increased heat flow across otherwise thermally insulating materials or constructions. They are a particular problem in building construction because it is rarely possible to provide a perfectly continuous and unbroken layer of insulation around the whole of the space which is to be insulated. In addition to the increased heat flow, a thermal bridge also has an effect on surface temperatures, creating in winter a drop in surface temperature on the inner warm side of the construction, and a rise in surface temperature on the external cold side. The lowering of the temperature on the inside brings with it the risk of surface condensation and mould growth, whereas the rise in temperature on the outside provides an opportunity for detecting the bridge by means of infrared thermography. The importance of a thermal bridge depends on a number of factors, including:

- The level of insulation which is required
- The area of the thermal bridge relative to that of the insulation material
- The thermal conductivity of the material of the bridge relative to that of the insulation material in which it occurs
- The geometry of the thermal bridge.

The three most common materials that create thermal bridges in building construction are timber, concrete and steel. When one of these materials forms a thermal bridge in an insulation layer, then, as the specified U-value is reduced, the thickness of insulation required to meet that U-value increases more and more rapidly. This effect is illustrated in Fig. 15.1, where insulation thickness is plotted against U-value for a typical construction element. In the lower curve, it is assumed that there is no thermal bridge in the construction, and that the insulation layer is perfectly continuous. In the upper curve, it is assumed that the insulation is bridged by 100 mm × 50 mm timber joists or battens. Note that the presence of the thermal bridge causes the necessary thickness of insulation to increase very rapidly as the U-value is reduced. Furthermore, the upper curve asymptotes to a minimum U-value, determined principally by the dimensions and properties of the thermal bridge, below which it is impossible to go.

This example demonstrates three things:

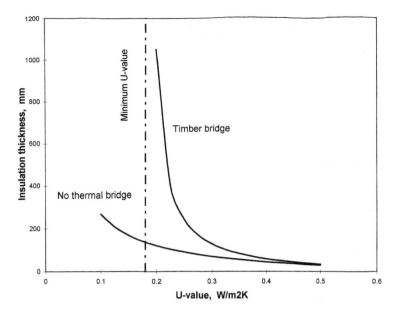

Fig. 15.1 The effect of a thermal bridge.

- At the U-values currently required for building elements, thermal bridges are highly significant
- Because of their significance, thermal bridges must either be avoided or be protected by thermal insulation
- Because a thermal bridge has such a large influence on the magnitude of the U-value of a construction element, the calculation procedure must be sufficiently sophisticated to take proper account of the bridge – otherwise the calculated U-value will be wrong.

The importance of adopting an appropriate calculation procedure for a construction element arises because the majority of simple methods tend to underestimate the effect of the bridge. A simple method will often give a result that is less than the correct U-value, giving the impression that the element is better than it really is. It also follows that when the U-value of a traditional construction element is re-calculated using currently recommended procedures, the new U-value is often found to be higher than the previously accepted value.

15.2 Thermal bridging, causes and avoidance

Thermal bridges in construction elements are due to a variety of causes, the most common being:

- Timber, concrete or steel columns, joists or battens in a wall or roof
- Lintels over openings

- Fixing and jointing details around the edge of openings
- Metal ties, staples, nails, etc. which penetrate an insulation layer, either to hold the insulation in place or to improve stability between the components on either side of the insulation layer
- Internal corners between walls, or walls and floors, or walls and ceilings.

The penetration of an insulation layer by services elements may also create thermal bridging. However, the creation of an air leakage path is likely to be a more serious problem.

In principle, avoidance is simply a matter of ensuring that all insulation layers are continuous and unbroken, and that all potential thermal bridging paths are either interrupted by at least some insulating material, or at worst made sufficiently small to be of little significance. In practice, avoidance is often difficult, and some possible solutions may not be consistent with other functional requirements such as, for example, strength and structural stability. Details designed to avoid or minimise thermal bridging can be found in a number of publications, such as the 'robust construction details' publication [2]. Manufacturers of construction systems are also publishing recommendations for their particular products, for example Kingspan [62]. In all cases, a recommended construction detail should include performance measures such as values for the linear thermal transmittance and the temperature ratio (these terms are defined in sections 15.3 and 15.5 below).

15.3 Calculation methods

Calculation methods for thermal bridges depend on the complexity of the bridge and the extent to which the bridge increases heat flow through the building fabric. Methods fall into one of three general categories.

15.3.1 Inclusion in the U-value

For bridges which are not too severe, the effect of the thermal bridge may be included in the calculation of the U-value of the element within which it occurs. This is the calculation procedure given in Appendix B of the Approved Documents and described in Chapter 5. It is satisfactory for some common types of thermal bridge, including:

- Repeated thermal bridges such as timber joists with insulation laid between, as in a roof
- Repeated thermal bridges such as timber battens with insulation fixed between, as in a timber frame wall
- Repeated thermal bridges such as mortar joints around blockwork
- Any other repeated thermal bridge where the components have similar thermal properties to any of the above.

The method also allows for corrections to be made for minor penetration of the insulation layer by, for example, metal wall ties and air gaps. When calculated in this way, the U-value includes the extra heat flow, and so no further account of the thermal bridge needs to be taken. The method is *not* suitable, and *cannot* be used for:

- Non-repeating thermal bridges
- Construction elements with metal connecting paths
- Construction elements with dense concrete columns, beams or lintels as the connecting path
- Corners and junctions in two or three dimensions
- Constructions with complex heat flow paths such as door and window frames.

15.3.2 Linear thermal bridges

Many thermal bridging effects in building structures may be treated as a linear thermal bridge. Examples are lintels (in any material), details around openings, junctions between plane elements (walls/roofs and walls/floors), or any similar detail that can be specified primarily in terms of its length. The extra heat loss due to the presence of the thermal bridge is accounted for by means of the linear thermal transmittance, Ψ, which is the heat loss coefficient per unit length of the bridge. The value of Ψ for a specific construction detail (such as a lintel or a wall/floor junction) may be found by laboratory measurement, but is more often obtained by calculation. Because the heat flow is two dimensional, the calculation procedure normally requires a numerical modelling method. Consider a detail, such as a lintel of length L and height y, as shown in Fig. 15.2.

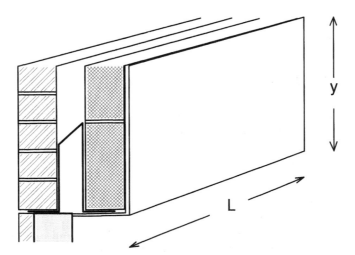

Fig. 15.2 Lintel detail.

The total fabric heat loss through this detail, including the bridging effect, may be written as:

$$H = L\ell^{2D}(T_i - T_o)$$

where ℓ^{2D} is the two dimensional heat transfer coefficient, determined from numerical modelling of the whole detail.

This heat loss may be considered in two parts: the heat loss that would have occurred if the bridge was not present, and the extra heat loss due to the presence of the bridge. The first of these is found from the U-value of the basic structure and its area, Ly. The second is found from the linear thermal transmittance:

$$H = L\ell^{2D}(T_i - T_o) = LyU(T_i - T_o) + L\Psi(T_i - T_o)$$

where   U is the U-value that the detail would have in the absence of the thermal bridge
        Ψ is the linear thermal transmittance of the detail.

Table 15.1   Maximum Ψ values.

Type of junction detail in external wall	Maximum Ψ,W/mK
Metal box lintel	0.30
Other lintels	0.21
Sill	0.04
Jamb	0.05
Ground floor	0.16
Intermediate floor within a dwelling	0.07
Intermediate floor between dwellings (apply half of the Ψ value to each dwelling)	0.14
Balcony within a dwelling (externally supported, not a continuation of floor slab, so that wall insulation is continuous and not bridged by balcony)	0.00
Balcony between dwellings (externally supported, not a continuation of floor slab, so that wall insulation is continuous and not bridged by balcony; apply half of the Ψ value to each dwelling)	0.04
Eaves (insulation at ceiling level)	0.06
Eaves (insulation at rafter level)	0.04
Gable (insulation at ceiling level)	0.24
Gable (insulation at rafter level)	0.04
Corner (normal)	0.09
Corner (inverted)	−0.09
Party wall between dwellings (apply half of the Ψ value to each dwelling)	0.06

Adapted with permission from BRE IP 17/01 [53]

Rearrangement of this equation provides the means for evaluating Ψ:

$$\Psi = \ell^{2D} - Uy$$

Further details on linear thermal bridges are given in BS EN ISO 10211-2: 2001 [18] and in IP 17/01 [55]. Typical maximum values of Ψ for several generic types of linear thermal bridge are given in Table 15.1.

15.3.3   Complex thermal bridges

Calculations for other more complex constructions normally require two or three dimensional numerical modelling methods. Specific techniques can be found in many fields of science and engineering, and methods of particular relevance to thermal bridges in construction elements are given in BS EN ISO 10211-1: 1996 [17]. The results of such calculations would, in many cases, be expressed in terms of an equivalent U-value; the calculations themselves are not likely to be part of normal building control assessment procedures.

15.4   Compliance with Part L

There are two ways of demonstrating compliance with the need to limit thermal bridging around openings and junctions. These are:

- By following the recommendations of the 'robust construction details' publication [2], and by adopting the design details published therein, either exactly or sufficiently closely as to achieve a performance that is just as good, *or*
- By demonstrating by calculation that the extra heat loss due to linear thermal bridges is within certain specified limits.

The first method requires no calculations. The second method is applicable when at least some of the details are not as specified in the 'robust construction details' publication. It is then necessary to know the values of the linear thermal transmittance for *all* relevant details in the external fabric. If *all* these values of Ψ are equal to or less than the values for their equivalent generic types in Table 15.1, only then may it be assumed that the requirement to limit thermal bridging has been satisfied. Otherwise it is necessary to proceed with a calculation.

The calculation procedure is based on writing the total fabric heat loss coefficient (i.e. the total fabric heat loss per unit temperature difference) as the sum of two terms:

$$\text{Total fabric heat loss coefficient, } H_T = \Sigma AU + \Sigma L\Psi$$

where   L is the length of the linear thermal bridge
Ψ is the linear thermal transmittance of the thermal bridge.

The $\Sigma L\Psi$ term includes the effect of all linear thermal bridges that have not otherwise been accounted for in the U-values of the ΣAU term. Ideally, the additional heat loss represented by $\Sigma L\Psi$ should be zero, but as this is almost impossible to achieve in practice, some leeway is allowed. The criterion to be met is that $\Sigma L\Psi$ should not exceed a certain proportion of ΣAU, when the latter is calculated using the appropriate maximum U-values specified for the elemental method (Tables 2.1 and 3.1). This criterion may be written:

$$\Sigma L\Psi \leq \alpha \, \Sigma AU_{\text{elemental}}$$

The coefficient α is chosen according to building type:

for dwellings $\alpha = 0.16$
for buildings other than dwellings $\alpha = 0.10$

The method by which this criterion is applied depends on the method which is being used to demonstrate compliance for the whole building. The following cases apply.

All buildings – elemental method

The criterion is the same for all buildings, and is exactly as given above:

For dwellings $\Sigma L\Psi \leq 0.16\Sigma AU_{\text{elemental}}$
For buildings other than dwellings $\Sigma L\Psi \leq 0.10\Sigma AU_{\text{elemental}}$

Dwellings – target U-value method

The total heat loss coefficient for a dwelling may be written:

$$H_T = A_T U_{AV} + \Sigma L\Psi$$

where A_T is the total area of all exposed elements, including the ground floor.

The criterion is formulated by using the target U-value for the dwelling instead of the elemental U-values of the component parts. Thus we require that H_T is within the limit given by:

$$H_T \leq A_T U_T + \alpha A_T U_T$$

Combining these two equations gives:

$$A_T U_{AV} + \Sigma L\Psi \leq A_T U_T + \alpha A_T U_T$$

Dividing by A_T allows the equation to be written as:

$$U_{AV} + \frac{\Sigma L \Psi}{A_T} \leq U_T + \alpha U_T$$

The left hand side of this equation may be considered as a modified average U-value, U'_{AV}:

$$U'_{AV} = U_{AV} + \frac{\Sigma L \Psi}{A_T}$$

The criterion may now be written, using $\alpha = 0.16$ for dwellings:

$$U'_{AV} \leq (1 + \alpha)U_T \quad \text{or} \quad U'_{AV} \leq 1.16 U_T$$

Dwellings – carbon index method

The fabric heat loss coefficient, H_T, appears in the SAP calculation procedure in box 37 of the SAP worksheet. If the heat loss via thermal bridges in the dwelling exceeds the allowable maximum, then the amount by which it exceeds this maximum must be added to H_T. This is done by adding a notional coefficient, $\Delta H_{notional}$, to the value of H_T in box 37, where:

$$\Delta H_{notional} = \Sigma L \Psi - \alpha \Sigma A U_{elemental} = \Sigma L \Psi - 0.16 \Sigma A U_{elemental}$$

The value of $\Delta H_{notional}$ found from this equation could in some circumstances be negative. If this were to happen, there is no guidance on whether or not it should be subtracted from H_T in box 37, or ignored.

Buildings other than dwellings – trade-off between construction elements

When applying trade-off, the requirement is that the transmission heat loss from the actual building is less than or equal to the heat loss for a notional building. This requirement may be written as:

$$H_{T\ actual} \leq H_{T\ notional}$$

The U-values in the notional building are taken as equal to the values in the elemental table, and so the allowance for thermal bridges is taken as a proportion of these. Thus:

$$H_{T\ actual} \leq (1 + \alpha)\Sigma A U_{elemental} \quad \text{or} \quad H_{T\ actual} \leq 1.1 \Sigma A U_{elemental}$$

Buildings other than dwellings – carbon emissions method

This is treated in a similar manner to the carbon index method for dwellings. The transmission heat loss coefficient, $H_{T\ notional}$, is evaluated for the notional building with which the actual building is being compared. This is then increased by $\Delta H_{notional}$:

$$\Delta H_{notional} = \Sigma L\Psi - \alpha\Sigma AU_{elemental} = \Sigma L\Psi - 0.1\Sigma AU_{elemental}$$

As with dwellings, there is no guidance on the procedure to be adopted should the calculated result for $\Delta H_{notional}$ be negative.

15.5   Condensation and mould growth

When the temperature of an inside surface is below the dew point of the air inside a building, condensation will occur. There are several possible causes of this condition, and the increased rate of heat loss due to a thermal bridge is one of them. Conditions are rarely so bad that condensation is a permanent feature on an internal surface, and in most cases, if it occurs at all, condensation appears intermittently. Because of this it can go unnoticed and may only become obvious when mould growth begins to disfigure the surface on which the condensation occurs. For an exposed element, the susceptibility of its inside surface to condensation and mould growth is related to the difference between the temperature of the internal environment and the temperature of the surface itself, expressed as a fraction of the overall temperature difference between the inside and outside environments. This ratio is a dimensionless temperature difference, and for steady state heat flow it depends only on the thermal properties of the structure. It may be defined as:

$$\text{Dimensionless temperature difference } \theta = \frac{T_i - T_{si}}{T_i - T_e}$$

where T_i is the temperature of the internal environment
        T_{si} is the temperature of the inside surface of an exposed element
        T_e is the temperature of the external environment.

Values of θ are often very small, and so an alternative dimensionless temperature difference, called the temperature factor, is preferred when setting criteria. This is defined as:

$$\text{Temperature factor }\quad f_{Rsi} = \frac{T_{si} - T_e}{T_i - T_e}$$

Note that $f_{Rsi} = 1 - \theta$

Defined in this way, when $T_{si} = T_i$, $f_{Rsi} = 1$, and surface condensation cannot occur. Thus the higher the value of f_{Rsi}, the less is the risk of condensation and mould growth. For the majority of buildings, if f_{Rsi} is 0.75 or higher for any exposed element, including a thermal bridge, neither condensation nor mould growth is likely to be a problem. However, in some buildings (or parts of buildings) the internal environment has a higher than normal humidity; examples are swimming pools, laundries, breweries, kitchens, canteens and spaces that use unflued gas heaters. In these cases f_{Rsi} should be at least 0.80, and preferably 0.90 or more.

The value of the temperature factor of a plane element with no thermal bridges can be calculated quite easily. A plane element, however, is not usually a problem unless its U-value is significantly higher than the maximum allowed by the elemental method. A thermal bridge is more likely to be susceptible to the risk of condensation and mould growth, and the temperature factor of a thermal bridge normally requires numerical analysis for its calculation. Values of the linear thermal transmittance and the temperature factor for specific details are often provided in manufacturers' technical literature.

15.6   Testing

Testing is normally carried out on the completed building. This makes it possible to detect thermal bridges that have arisen due to both faulty components and errors during construction. The majority of test methods involve the measurement of the surface temperature of the building, and the most convenient and universally applied technique is infrared thermography.

15.6.1   Infrared radiation

All matter continuously emits and absorbs electro-magnetic radiation. An increase in the surface temperature of a material causes an increase in the intensity of radiation emission, and also alters the wavelength distribution of that radiation. Perfect emitters, that is materials which emit the maximum possible intensity of radiation at all wavelengths, are called 'black body' radiators. All normal materials emit less than a black body, the difference being described by the emissivity of the material, which is defined as the ratio of the amount actually emitted to the amount that would be emitted by a black body at the same temperature. The emissivity of a material usually varies with wavelength and is often low at short wavelengths, such as in the visible region of the spectrum, and high in the infrared region. For some materials, this variation with wavelength is rapid and erratic, but for most of the natural materials used in building the emissivity is relatively constant, especially in the infrared region, where it usually has a high value of 0.8 or more. Metals and materials with a metallic finish are the most likely to have an emissivity which varies strongly with wavelength, but as used in construction, metals normally have a non-metallic surface coating, even if it is only an oxide of the metal which has been deposited by the normal processes of weathering. The fact that most materials have stable radiation properties in the infrared region of the spectrum, and emit large amounts of radiation because their emissivity is high, makes it possible to utilise measurement of that radiation to determine their surface temperature.

15.6.2   Infrared thermography

Infrared thermography uses specially designed infrared video or still cameras that convert the infrared radiation into images that show surface heat variations. The technique has applications in a variety of fields, including medicine, electrical installations, cold storage facilities, etc. When applied to buildings, it can be used to examine the whole or part of the exterior envelope, or it can be used to look at surface temperatures within a building. When used to identify thermal bridges, it is usual to take images of the exterior of the building. Such images, when combined with an air pressurisation test, can also indicate points at which air is leaking from the building.

However, although thermographic scans of the outside of a building are more convenient, especially if the building is large, they have a number of drawbacks. Warm air escaping from a building does not always move through the walls in a straight line, and heat loss detected in one area of an outside wall might originate at some other hard-to-find location inside the wall. Wind conditions also affect the thermal image. On windy days, it is harder to detect temperature differences on the outside surface of the building. The most accurate thermographic images usually occur in calm conditions when there is a large temperature difference (at least 14°C) between inside and outside air temperatures. For this reason, in the UK, thermographic scans are most effective when done in winter. Unfortunately, even in ideal conditions, the temperature differences between different parts of the exterior surfaces of a building are not large, and the thermal imaging equipment must have a very high resolution in order to detect them.

Further information on the application of infrared thermography to building inspections is given by Hart [63].

References

BRE   Building Research Establishment, Watford, Herts WD25 9XX

BRECSU   Building Research Energy Conservation & Support Unit, Watford, Herts WD25 9XX

CRC   Construction Research Communications Ltd, Watford, Herts WD25 9XX

CIBSE   Chartered Institution of Building Services Engineers, 222 Balham High Road, London SW12 9BS

1. *Thermal insulation: Avoiding risks.* BR 262, BRE/CRC, 2002.
2. *Limiting thermal bridging and air leakage: Robust construction details for dwellings and similar buildings.* The Stationery Office, 2001.
3. *Building materials and products – Hygrothermal properties – Tabulated design values.* BS EN 12524: 2000.
4. *Guide A: Environmental design, Section A3: Thermal properties of building structures.* CIBSE, 1999.
5. *Thermal performance of building materials and products – Determination of thermal resistance by means of guarded hot plate and heat flow meter methods – Dry and moist products of low and medium thermal resistance.* BS EN 12664: 2001.
6. *Thermal performance of building materials and products – Determination of thermal resistance by means of guarded hot plate and heat flow meter methods – Products of high and medium thermal resistance.* BS EN 12667: 2001.
7. *Thermal performance of building materials and products – Determination of thermal resistance by means of guarded hot plate and heat flow meter methods – Thick products of high and medium thermal resistance.* BS EN 12939: 2001.
8. *Thermal insulation – Determination of steady-state thermal transmission properties – Calibrated and guarded hot box.* BS EN ISO 8990: 1996.
9. *Thermal performance of windows and doors – Determination of thermal transmittance by hot box method – Part 1: Complete windows and doors.* BS EN ISO 12567-1: 2000.
10. *Thermal performance of buildings – Transmission loss coefficient – Calculation method.* BS EN ISO 13789: 1999.
11. *Conventions for the calculation of U-values.* BRE/CRC, 2002.
12. *Building components and building elements – Thermal resistance and thermal transmittance – Calculation method.* BS EN ISO 6946: 1997.
13. *Thermal performance of buildings – Heat transfer via the ground – Calculation methods.* BS EN ISO 13370: 1998.
14. *Thermal performance of windows, doors and shutters – Calculation of thermal transmittance – Part 1: Simplified methods.* BS EN ISO 10077-1: 2000.

15. *Thermal performance of windows, doors and shutters – Calculation of thermal transmittance – Part 2: Numerical method for frames.* prEN ISO 10077-2.
16. *Basements for dwellings.* Approved Document, ISBN 0-7210-1508-5, British Cement Association and National House Building Council, 1997.
17. *Thermal bridges in building construction – Calculation of heat flows and surface temperatures – Part 1: General methods.* BS EN ISO 10211-1: 1996.
18. *Thermal bridges in building construction – Calculation of heat flows and surface temperatures – Part 2: Linear thermal bridges.* BS EN ISO 10211-2: 2001.
19. *SAP: The Government's Standard Assessment Procedure for energy rating of dwellings.* BRECSU, 2001. (The most up-to-date edition is available at www.bre.co.uk)
20. *Lighting for buildings: Code of practice for daylighting.* BS 8206: 1992; Part 2
21. *Testing buildings for air leakage.* TM23, CIBSE, 2000.
22. *Copper indirect cylinders for domestic purposes – Specification for double feed indirect cylinders.* BS 1566-1: 1984.
23. *Specification for copper direct cylinders for domestic purposes.* BS 699: 1884.
24. *Specification for copper hot water storage combination units for domestic purposes.* BS 3198: 1981.
25. *Specification for unvented hot water storage units and packages.* BS 7206: 1990.
26. *Performance specification for thermal stores.* Waterheater Manufacturers Association, 1999.
27. *Controls for domestic central heating and hot water systems.* GPG 302, BRECSU, 2001.
28. *Specification for installation in domestic premises of gas-fired ducted air heaters of rated output not exceeding 60 kW.* BS 5864: 1989.
29. *Method for specifying thermal insulating materials for pipes, tanks, vessels, ductwork and equipment operating within the temperature range −40°C to +700°C.* BS 5422: 2001.
30. *Low energy domestic lighting.* GIL 20, BRECSU, 1995.
31. HETAS Limited, PO Box 37, Cheltenham GL52 9TB.
32. *Energy efficient refurbishment of existing housing.* GPG 155, BRECSU.
33. *Planning and the historic environment.* Planning Policy Guidance PPG15, DoE/DNH, September 1994. (In Wales, refer to *Planning guidance Wales planning policy first revision* 1999, and *Planning and historic environment: Historic buildings and conservation areas*, Welsh Office Circular 61/96.)
34. *The principles of the conservation of historic buildings.* BS 7913: 1998.
35. Information Sheet 4: *The need for old buildings to breathe.* Society for the Protection of Ancient Buildings, 1986.
36. *Energy efficiency in buildings.* CIBSE, 1999.
37. Technical Note 14: *Guidance for the design of metal cladding and roofing to comply with Approved Document L.* Metal Cladding and Roofing Manufacturers Association, 2002.
38. Guide A: *Environmental design,* section A5: *Thermal response and plant sizing.* CIBSE, 1999.
39. CHPQA Standard: *Quality assurance for combined heat and power, issue 1.* DETR, November 2000.
40. *The designer's guide to energy-efficient buildings for industry.* GPG 303, BRECSU, 2000.

41. *Heating controls in small commercial and multi-residential buildings.* GPG 132, BRECSU, 2001.

42. Guide H: *Building control systems.* CIBSE, 2000.

43. *Photoelectric control of lighting: Design, set-up and installation issues.* IP 2, BRE/CRC, 1999.

44. *The Carbon Performance Rating for offices.* Digest 457, BRE/CRC, 2001.

45. *Energy assessment and reporting methodology: Office assessment method.* TM22, CIBSE, 1999.

46. Variable flow control. GIR 41, BRECSU, 1996.

47. *Guidance for environmental design in schools.* Building Bulletin 87, The Stationery Office, 1997.

48. *Achieving energy efficiency in new hospitals.* NHS Estates, The Stationery Office, 1994.

49. *Building energy and environmental modelling.* AM11, CIBSE, 1998.

50. *A practical guide to infra-red thermography for building surveys.* BR 176, BRE/CRC, 1991.

51. Commissioning Code A: *Air distribution systems* (1996), Commissioning Code B: *Boiler plant* (1975), Commissioning Code C: *Automatic controls* (2000), Commissioning Code R: *Refrigerating systems* (1991), Commissioning Code W: *Water distribution systems* (1994). CIBSE, 1975–2000.

52. Technical memorandum 1: *Standard specification for the commissioning of mechanical engineering services installations for buildings.* Commissioning Specialists Association, 1999.

53. *Sub-metering new build non-domestic buildings: A guide to help designers meet Part L of the Building Regulations.* GIL 65, BRECSU, 2001.

54. Energy consumption guide 19 (ECON 19): *Energy use in offices.* DETR, 1998.

55. *Assessing the effects of thermal bridging at junctions and around openings.* IP 17/01, BRE/CRC, 2001.

56. Etheridge, D. & Sandberg, M., *Building Ventilation Theory and Measurement.* Wiley, 1996.

57. Potter, N., *Air tightness testing.* Technical Note 19, Building Services Research & Information Association, 2001.

58. Perera, M.D.A.E.S., Turner, C.H.C. & Scivyer, C.R., *Minmising air infiltration in office buildings.* Building Research Establishment Report, BRE/CRC, 1994.

59. *Performance of windows: Part 1: Classification for weather tightness.* BS 6375: 1989.

60. Liddament, M.W., *A guide to energy efficient ventilation.* Air Infiltration and Ventilation Centre, 1996.

61. Serhman, M.H. & Grimsrud, D.T., *Measurement of infiltration using fan pressurisation and weather data.* Proceedings of AIC Conference, *Instrumentation and Measuring Techniques,* 1980.

62. *Approved construction details for Part L2 (England and Wales) and Part J (Scotland).* Kingspan Insulated Panels, Greenfield Business Pk2, Holywell, Cheshire CH8 7GJ

63. Hart, J. *An introduction to infra-red thermography for building surveys.* IP 7/90, BRE/CRC, 1990.

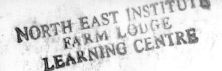

Other Sources of Information

Avoiding or minimising the use of air conditioning. GIR 31, BRECSU, 1995.

Metal cladding: Assessing thermal performance. IP 5/98, BRE/CRC, 1998.

U-value calculation procedure for light steel frame walls. BRE/CRC, 2002.

Solar shading of buildings. BR 364, BRE/CRC, 1999.

Central heating specifications (CHeSS). GIL 59, BRECSU, 2000

Guide to good practice for assessing glazing frame U-values. Centre for Window and Cladding Technology, 1998.

Guide to good practice for assessing heat transfer and condensation risk for a curtain wall. Centre for Window and Cladding Technology, 1998.

Guide for assessment of the thermal performance of aluminium curtain wall framing. Council for Aluminium in Building, 1996.